D0093385

mjf

THE COSMIC POWER WITHIN YOU

JOSEPH MURPHY
D.R.S., Ph.D., D.D., LL.D.

MJF BOOKS
NEW YORK

Published by MJF Books
Fine Communications
Two Lincoln Square
60 West 66th Street
New York, NY 10023

Library of Congress Catalog Card Number 95-82185
ISBN 1-56731-108-3

This edition is reprinted by arrangement with Prentice-Hall Inc./Career & Personal Development.

Manufactured in the United States of America

10 9 8 7 6 5 4 3 2 1

What This Book Can Do For You

Whatever you desire in life you may have, for there is a Cosmic Power within YOU which can bring all your dreams to fruition. *This Cosmic Power is the greatest force in all the world.* You can begin to use this Power *now* for healing your body, and for prosperity in your business or professional world. This Cosmic Power will guide and direct you, attract to you the right companions or partner, bring out your hidden talents, and help you prosper in countless ways. It is the same Power and Principle by which inventions, securities, automobiles, airplanes, or real estate are created and successfully sold.

Many books have been written about the powers of the mind, but *in this book you are told explicitly how to use this Power to transform your whole life,* how to think and image constructively and successfully, how to enter into the life more abundant, and how to achieve peace, satisfaction, and serenity in this changing world.

You have unlimited creative possibilities, and every chapter of this book teaches you how to tap this Cosmic Power in order to get the most satisfying results for a richer, fuller, and more creative life. As you learn in this book the simple processes and techniques for contacting the Cosmic Power and begin to apply this wonder-working Power in your life, you will move onward, upward, and Godward.

I have taught thousands of people all over the world how to use this Cosmic Power to draw love into their lives, and bring harmony where discord is, peace where pain is, joy where sadness is, health where sickness is, wealth where poverty is.

You Can Experience an Exciting Adventure

You and I are about to explore and plumb the storehouse of wisdom, power, and all the treasures of the Cosmic Power within us. You will learn how to use this Cosmic Power in your daily life, in your personal relationships, in the healing of marital problems or discord in your home or office, and in all other phases of your life. You will learn how to contact and to use this miracle-working Power to bring countless blessings into your life and into the lives of others.

You Will Learn from the Experiences of Other People

In this book you will read about men and women in science, art, business, and industry who have used this Cosmic Power. *They tell you exactly how they used It* to bring about health, happiness, and success, as well as the realization of their hearts' greatest desires. Some have given me express permission to use their letters, including their names and addresses; in other cases, I have used fictitious initials to guard the identity of the writers.

Follow the techniques outlined step by step, and, like all those described in the book, you, too, will get results beyond your fondest dreams.

The Purpose of This Book

This book is clearly designed to reveal to you in a simple, down-to-earth manner and in everyday language how to lead a richer, fuller, and more glorious life. All you have to do is to use the

Cosmic Power within you, which is always available, waiting for you to call upon It.

Cease looking outside. Look inside yourself and make the magic contact, for as you change your attitude and mind, you change your world. Within this book you will find the key to successful and triumphant living as you want to live it.

The techniques and programs presented in this book will enable you to use the mental and spiritual laws on which I have been writing and lecturing for the past thirty years. The methods and processes presented in the pages of this book have already helped thousands of people to experience guidance, health, prosperity, happiness, and peace of mind.

You are exactly what you think all day long. Consequently you are the artificer of your own future. As you change your thought-life or pattern, you change your destiny. The study and application of the great Cosmic truths outlined in each chapter will result in the most fruitful, rewarding, and richest experiences you have ever had.

Begin now to command your life pattern and go forward to achievement, accomplishment, and victory, and experience the life more abundant—right *here* and right *now!*

Contents

1. HOW TO KEEP IN TUNE WITH THE COSMIC POWER 1

How a paralyzed arm was healed—Your infinite reservoir of power—Constant renewal and refreshment always available—You are never out of touch with Cosmic Guidance—How to tune in for a safe journey—How a college student tuned in for passing examinations—How a long-lost brother was found—How a widow healed her grief—How a businessman tunes in to Cosmic Power—The ideal way to tune in—Summary

2. HOW YOUR COSMIC SUBCONSCIOUS CAN GUIDE YOU 14

Your subconscious can guide you without error—The source of Gandhi's tremendous spiritual power—How a diplomat's wife increased her graciousness—Hidden talents found through Infinite Intelligence—The right way to practice Divine Guidance—Solving an "impossible" situation—Become the master of every situation—Getting ahead using Cosmic Truths—Use the Cosmic Power and win outstanding victories—Learn to be yourself—Thinking makes it so—How to gain faith yielding real benefits—How faith in the God-Presence overcame failures—How to become friendly, happy, joyous and free—Summary

3. HOW TO BECOME AWARE OF YOUR COSMIC POWER 25

The powers within — Self-esteem gained through Cosmic awareness — Shyness and timidity overcome by using the Cosmic Power — The true meaning of self-love — Learn to love yourself — Self-condemnation and annoyance overcome — Practicing the Golden Rule — Look at the small end — How to get a higher estimate of yourself — Health improved with new self-appraisal — Formula for business success — Summary

4. HOW THE COSMIC POWER CAN SOLVE PROBLEMS 38

Ninety per cent of problems are humanly created — The right way to solve problems — How changed thinking healed ulcers and high blood pressure — How the Cosmic Power effected a promotion — How a mother's mental movie worked wonders — A healing of personality for business success — How you can develop a marvelous personality — How a secretary practiced empathy — How a silence between sisters was broken — His application of the Golden Rule resulted in a wonderful promotion — You are needed regardless of your age — Saying "yes" to life in Cosmic Truth — How you can triumph over depression and all obstacles — Summary

5. HOW TO USE THE COSMIC HEALING POWER . 50

How a crippled hand was healed — How Cosmic Power healed a diseased kidney and a broken bone — The Cosmic Healing Power and how you can use it — How the Cosmic Power healed a tailor's blindness — Twentieth-Century miracles of healing — How the Cosmic Power resolved a writer's manuscript problem — How a schoolteacher healed her ulcers and achieved promotion — How a five-word formula healed epilepsy — The law of belief and how to use it — Summary

6. HOW TO LEAD A SUCCESSFUL LIFE AND GAIN PROMOTION 59

How a woman received specific guidance — A hidden talent revealed — Accept health, wealth, and happiness NOW — How to plan a glorious future for yourself — How a maid got a wonderful car by writing a letter to herself — How an eight-year-old boy received a gift he wanted — How Cosmic Love responded to a widow — How a promotion and an enormous increase in salary were obtained — How a new lease on life was secured — How to experience happiness and success in your life right now — The joy of overcoming all your problems with Cosmic Science — Summary

7. THE GREATEST SECRET OF THE AGES . . . 72

The truly effective prayer which transforms your life — Keep your eyes on the Cosmic beam and move ahead in life — How a space scientist gets answers in space research — Get a new image and become what you want to be — How you can achieve great things — How to banish a "jinx" — Her belief in Cosmic Power healed her — Let your head-knowledge become heart-knowledge — Why doesn't God do something about war, crime and disease? — The superstitious origin of the belief in two powers — Your freedom to choose health, happiness, and prosperity — The knowledge that gives you peace, harmony, and answers to your problems — Good and evil determined by your thought — You become what you contemplate — Summary

8. HOW TO MAKE RIGHT DECISIONS WITH COSMIC POWER 85

The power of decision — How his power of decision

won him a new car — How courage to decide transformed a life — How power of decision brought about a miraculous healing — How a pharmacist makes right decisions — An effective prayer for right decision — Logical decisions will guide you — How a woman became a stockbroker — Decide to accept your Cosmic Divinity now — The Cosmic Power is no respecter of persons — What happens from lack of decision — The Cosmic Power backs up all your decisions — How an alcoholic was healed through his power of decision — Summary

9. COSMIC POWER — YOU MOST POWERFUL FRIEND 96

Cosmic Power used to heal a dying son — How an educator used Cosmic Power to vanquish all obstacles — How his Cosmic "partner" saved him $250,000 — How Cosmic Power saved a life through a dream — Subconscious instinct — How the Cosmic Mind forms habits and reveals answers — How immediate guidance was received from Cosmic Mind — Expect an answer — Summary

10. THE COSMIC VISION OF "HEALTHY-MINDEDNESS" 107

How Cosmic Vision paid fabulous dividends in health, wealth, and happiness — How the Cosmic Power revealed a life-saving idea — How the Cosmic Power healed a cripple — The miracle of faith and love — How a mother's vision worked miracles for her son — You can rise in new confidence through the Cosmic Power within you — Summary

11. THE MAGIC OF FAITH IN THE COSMIC POWER 116

Everyone has faith in something — How faith paid off a $25,000 mortgage — How to increase your faith in your power to realize your desire — How a playwright used the magic of his new faith in the Cosmic Power — Self-doubt and fear conquered through faith in the Cosmic Power — A health miracle wrought — Within you is the Cosmic Power to work wonders in your life — A minister's acute migrain attacks healed — Summary

12. HOW TO GET WHAT YOU GO AFTER 127

You create your own destiny — How a tremendous increase in business was achieved — He stopped blaming fate and achieved promotion — How a salesman broke blocks to closing sales — What you can conceive, you can achieve — Summary

13. HOW TO BE JOYOUS CONQUERING OBSTACLES TO YOUR COSMIC GOOD 137

How Cosmic Mind helped sell real estate — How a businesswoman triumphed over obstacles — The answer that saved the suicide — How he found a way out of the wilderness in Vietnam — How an executive overcame pressure and excess tension — Summary

14. THE COSMIC POWER AND YOUR FUTURE . . 147

A new future gained through Cosmic Healing Power — How to assume responsibility for the future — It's never too late to change for the better — Why you are always planning for the future — How a young woman discovered she has what it takes to face life — How a salesman changed his mind and changed his destiny — Summary

15. HOW TO BE SERENE IN A CHANGING WORLD 158

How an alcoholic found inner peace and freedom — She learned to forget bitterness and found serenity —

How an executive found true serenity—How you can find complete serenity in any circumstance—The prayer that gives you serenity in this changing world—Summary

16. HOW TO OVERCOME WORRY 167

How business worry was overcome—How a mother dispelled her inner fears—A practical and workable approach to overcoming anxiety—How fear of auto travel was conquered—Your invisible partner—How a schoolteacher disposed of worries—Emotional spasms healed—How worry of high blood pressure was handled—A prayer to banish chronic worry—Summary

17. HOW TO LINK YOUR THOUGHT TO COSMIC POWER 178

How thinking in Cosmic Truth brought promotions—A remarkable case of a man from India—How to know if you are really thinking with Cosmic Power—How a mother stopped reacting to negative suggestions—A case of a seemingly virtuous man deprived of good fortune—Beat the law of averages—Summary

18. LET THE COSMIC POWER WORK WONDERS FOR YOU 191

How a wife reclaimed her husband's love—He said, "This is my fifth divorce! What's wrong?"—Why a manuscript was not published—She believed the "cards were stacked" against her happiness—He prayed and worked hard, but didn't prosper—He wanted to outlaw war by passing a law—Why doesn't God stop wars if God is love?—Summary

1

How to keep in tune with the Cosmic Power

There is a Power in you that has never been fully released. The Power that moves the world and governs the galaxies in space is *within you.*

I have known people to tune in with this amazing, untapped Power within them, and in a short time their whole lives were transformed so radically that their friends barely recognized them. Many of them received this kind of greeting from their old acquaintances: "I wouldn't have known you! What happened to you?"

You possess limitless possibilities for development, and as you begin to tune in on the Cosmic Power within you, you will discover that this inner Power can lift you up from sickness, failure, misery, despondency, or utter chaos. It can heal your body, guide and direct you, open up new doors for you, dry your tears, dissolve your problems, and set you on the high road to happiness, freedom, and peace of mind.

This Cosmic Power can inspire you. If you call upon IT, you will receive an answer. If you are seeking your true place in life, you may call upon this primal Power and IT will open up a new door for you and guide you every step of the way. It is your right to keep in tune with this Power and to allow IT to flow through your mind, body, and affairs, enabling you to move onward, upward, and Godward along all lines.

How a Paralyzed Arm Was Healed

The following letter reveals how a woman who attends my lectures contacted this wonderful Power:

> Dear Dr. Murphy: Words cannot express my gratitude for your instructions on how to contact the Infinite Power within me and the marvelous response I received. I began to think about what you said—that all the power of the Godhead was within my own mind and that I could use the Power. I began to think that this Power was greater than hydrogen bombs, atomic energy, or electricity, and that all these were as nothing compared to the all-powerful God within me.
>
> For ten days I had been unable to lift my arm upward without unbearable pain, which caused me to scream at times. On the way to my family doctor, I tuned in on the Cosmic Power and affirmed: *Through the power of the Almighty within me, I am now moving my arm freely.* I lifted my arm to a horizontal position with no pain whatsoever, and I moved it more freely than had been possible for some weeks. The doctor tested my arm, and all goes well. Truly, the Kingdom of God *is* within each one of us! Respectfully, Mrs. Helen Hanford, Los Angeles, Calif.

Mrs. Hanford consciously discovered this Power within her own mentality, and a marvelous healing resulted. Attach yourself to God's power and *let wonders happen* in your life!

Your Infinite Reservoir of Power

The Power within you is inexhaustible, eternal, and infinite. You have reserves of infinite wisdom, boundless love, infinite intelligence, absolute harmony, absolute peace, fullness of joy, indescribable beauty and the infinite healing Presence—all these powers, capabilities, and energies are within you waiting to be called upon.

A general operating a military force has at hand a reserve of men and equipment on which he can call; likewise, when you are perplexed, confused, muddled, fearful, or depressed, you can call on your spiritual reserves and receive spiritual refreshment and be replenished with new wisdom, truth, and beauty.

Constant Renewal and Refreshment Always Available

The following letter points out how one person constantly draws from the treasure-house of Infinity within her:

> Dear Dr. Murphy: Just a year ago, a friend of mine took me to one of your lectures on a Sunday morning. I had just separated from my husband and was filing for divorce after 18 years of a difficult marriage. I was depressed, confused, frightened, and full of guilt feelings. I am now constantly in tune with the Infinite, and miracles are happening in my life.
>
> Your lecture that morning really "hit home," and made me realize I must learn to control my thinking and my emotions. Before, I was letting family, friends, and negative thoughts of the world control me.
>
> Since then, with the help of your lectures and a number of your books to guide me, marvelous things have happened and are happening to me. Gone are the nerve pills and the migraine headaches which were a part of my daily routine.
>
> At 40, I am more radiant, healthy, and happy than I have ever been in my entire life, and my calm, positive, healthy outlook on life has made a great difference in the lives of my two teen-aged boys. My blessings are abundant and I give thanks every moment of every day. New doors are opening, and greater prosperity and peace of mind are mine.
>
> I have not completely conquered. Once in a while, for no apparent reason, I feel myself slipping into the old ways of thinking or feeling but, thanks to what I have learned from you, I am now able to sit quietly, withdraw into myself, and come out of it by calling on God, and His peace

and love flow in response. Many, many thanks again for opening my eyes to the glory surrounding me. Sincerely, Mrs. E.C., Los Angeles, California.

This woman's constant prayer is as follows:

God is, and His Presence flows through me as harmony, joy, peace, beauty, and right action. His love fills my soul, and wonders happen as I pray.

She uses this prayer as a sort of melody and sings it to herself as she goes about her household and other duties. Just as a general in the field keeps in contact with the Pentagon in Washington, D.C., for instructions, so also does this wonderful woman receive instruction, guidance, and Divine ideas by keeping in contact with God and Cosmic Wisdom, and she is furnished with prosperity that means freedom to be, to do, and to go as her mind dictates.

You Are Never out of Touch with Cosmic Guidance

The following letter shows how, when you take a trip, you can keep in touch with the source of strength, peace, and security that is constantly available:

Dear Dr. Murphy: .... To say that I was grateful is such a feeble expression of my feelings! How well I know that when you turn to God as I did, He will turn to you in love. You are never alone when you know, as I did on this trip, that *God did go before me* to make perfect my way.

I do not have to return to Monterey for my final five-year X-ray until December, as is done in all cases of previous cancer, but I have no fear, for I know that I am permanently cured for all time and that this disease which so many people call incurable is not—but that there are incurable people.

I took your wonderful and loving letter written to me on July 12th ... in which you wrote, "God is in His holy temple and, that being true, the Presence flows through you as beauty, love, harmony, and peace, and in your flesh you shall see God." Well, I carried that letter with me and read and re-read it many times a day while I was away.

If there is any portion of this letter which you wish to use in any way in an effort to help others, you are fully free to do so. I so wish to help any and all who may need my assurance.

My heart, you know, is filled with gratitude, and as you said this morning on the radio, "This is the day the Lord hath made; be happy and rejoice therein." Gratefully, Mrs. R.

How to Tune in for a Safe Journey

You can keep in constant contact with the boundless wisdom and Infinite Power which transcend our feeble intellect, when going on a journey by plane, automobile, train, or by any other means of conveyance. Affirm feelingly and knowingly:

This vehicle is God's idea moving from point to point freely, joyously, and lovingly. Divine love goes before me, making straight, beautiful, joyous, and happy my way. The sacred circle of God's eternal love surrounds me, enfolds me, and enwraps me, and I am always in the midst of the Holy Omnipresence. I am always surrounded by the whole armor of God and Good, and it is wonderful!

How a College Student Tuned in for Passing Examinations

Dear Dr. Murphy: As you know, I have missed several classes because of illness, but I have been praying that God would supply all my needs for the examination and reveal to me what I need to know, as you suggested. Last Monday morning the examination took place, and on Sunday night I had a dream in which you appeared and pointed out the pages to study in my physics and chemistry books. I arose at 2 AM and studied the pages outlined in my dream. One page I committed to memory.

Needless to say, I was able to answer all the questions, as I had read over and over the answers in my two books the night before. I told one of the professors about my dream, and he laughed at me and thought I was joking! Many thanks for your help. D.L., Beverly Hills, California.

Infinite Intelligence within the subconscious mind of this boy responded to his call, as Its nature is responsive. This same Infinite Intelligence directed the professors to ask the same questions as the boy saw in his dream. The Bible says: . . . *I, the Lord* (the law of your Cosmic subconscious) *will make myself known unto him in a vision, and will speak unto him in a dream* (Numbers 12:6).

How a Long-Lost Brother Was Found

The Bible says: *Thou wilt keep him in perfect peace, whose mind is stayed on thee: because he trusteth in thee* (Isaiah 26:3).

A man wrote me saying that he had not seen his brother in over twenty years and did not know where he was. In the meantime an estate was bequeathed to both of them, and he was anxious to communicate the good news to his brother. His letter reads as follows:

Dear Dr. Murphy: I read your book, *The Miracle of Mind Dynamics,*[1] and was very much impressed. I kept my mind focussed on Infinite Intelligence; I could not see It, but based on previous experiences I was convinced of Its reality. I don't see the wind, but I feel its breeze upon my face.

I asked this Infinite Intelligence to reveal the whereabouts of my brother and I kept repeating, *Divine guidance is mine now, and Infinite Intelligence brings us together.* Last week I attended a conference in New York, and lo and behold! one of the passengers who sat next to me on the subway was my brother whom I have not seen for over twenty years!

[1] *The Miracle of Mind Dynamics,* Joseph Murphy, Prentice-Hall, Inc., Englewood Cliffs, New Jersey, 1964.

> I thought you might like to use this letter for the new book you are writing. I am sure this experience will instill in all of us faith and confidence in the mystic Power. Signed, T.L., San Francisco, California.

You must realize that there is an infinite ocean of life operating in all of us, filled with adequacies and answers to all problems. Ralph Waldo Emerson said, "There is guidance for each of us, and by lowly listening, *we shall hear the right word.*"

Action and reaction are cosmic and universal. You are dealing with a reciprocal action-and-reaction universe: as you sow, you reap; and as you call, you receive an answer.

How a Widow Healed Her Grief

The following letter lucidly demonstrates how, in time of grief and tragedy, you may tune in on the river of peace within you and receive a sense of inner tranquility, poise, balance, and serenity.

> Dear Dr. Murphy: Following the death of my beloved husband, I was depressed and discontented. A friend invited me to hear you speak at the Wilshire Ebell Theatre. You said that it is natural to miss our loved ones and that a good cry is a part of the healing process and not to repress or to suppress the tears at the time of a transition. You also said that protracted grief is wrong, as it robs one of vitality, enthusiasm, and energy, and that the answer is prayer for the loved one who is now living in the next dimension of mind. What really healed me was your explanation of how our loved ones are all around us, separated only by frequency. Just as voices are sent over a cable at different frequencies or as, when a fan rotates at a very high speed it becomes "invisible" and you can see a wall through the blades, so are our beloved ones separated from us.
>
> Suddenly during your lecture I realized that my late husband was as much alive as I was, that the reality of him was mind, spirit, or consciousness, and that his body was just an instrument and he now had another body, in a more rarified and attenuated form which is called fourth-dimensional.

> I realized there is no death because Life was never born and will never die. I began to pray for my husband as you outlined; the following was my prayer:
>
> *I surrender my husband to God. I know he is alive with the life of God and that his journey is ever onward, upward, and Godward, as life goes not backward nor tarries with yesterday. I radiate love, peace, joy, and goodwill to him, and I know that God's love and peace surround him and enfold him. The Light of God shines in him, through him, and all around him. I know goodness and mercy follow him all the days of his life and he dwells in the house of God forever. His journey is from glory to glory, and whenever I think of him I say, God be with you.*
>
> Having prayed this way for a few days, a great sense of peace came over me and suddenly I knew inwardly that life is eternal and love is immortal. Truly, God wiped away my tears and there was no more crying. Lovingly, Mrs. X.

Tuning in with the Cosmic Power and experiencing its bliss, harmony, peace, and joy is the answer to all problems. The Bible says, *Thou wilt keep him in perfect peace, whose mind is stayed on thee. . . .* (Isaiah 26:3) The word "thee" is interpreted to mean the Presence of God or Infinite Life within you.

The ancient Hindu scriptures said of this Life-Principle which animates all men and is their very Reality: "It was never born and it will never die; water wets it not; fire burns it not: wind blows it not away. Knowing these things, why grievest thou for it?" Our Bible says, . . . *Why seek ye the living among the dead? He is not here, but is risen* (Luke 24:5-6). Your loved one is closer to you than ever before.

The Bible explains eternal life beautifully in John 17:3: *This is life eternal, that they might know thee the only true God. . . .*

How a Businessman Tunes in to Cosmic Power

Herewith is a letter which shows the wonders of regularly tuning in with the Cosmic Power within you:

> Dear Dr. Murphy: I have read in the Bible that faith is the substance of things hoped for, the evidence of things not

seen.[2] I know that my faith is an actual state of mind, the evidence and sure forerunner of prosperity and success in life. My faith is based on the laws of my mind as mentioned in your book, *The Power of Your Subconscious Mind.*[3] I now know that the law of my subconscious mind, properly used, can't fail me. I used to have a habit of saying, when I came down to breakfast, "This is going to be another one of those black days. I hate to go to work! I can't stand the boss and the country is going to the dogs"—and much more of this negative thinking.

Last January, after reading *The Miracle of Mind Dynamics* and *The Amazing Laws of Cosmic Mind Power,*[4] I started a procedure of affirming certain truths every morning after I arrive at the office. I remain absolutely quiet and then, after I look at my schedule for the day, I ask my secretary not to disturb me for fifteen minutes. Then I pray for peace, harmony, and guidance for the day. I use the following prayer every morning:

God dwells at the center of my being. God is peace; this inner river of peace enfolds me now. I act with confidence and enthusiasm all day long. All my employees in the office and the field are Divinely guided and prospered in all their ways. There is Divine understanding between myself and all my associates and employees. There is a Divine, harmonious union of our minds and hearts, and our business and their families are blessed and prospered.

The love and light of God watch over all of us like a loving mother watches over her sleeping child. My decisions are Divinely directed. When I hold a conference I am inspired to say the right thing which blesses all. I draw continually from the Divine Storehouse, and I know that my faith in God and His laws is immediately translated into health, money, success, love, and harmony, which I and my associates and employees seek right here and now. I radiate love, peace, and goodwill to all and I am at peace.

I have followed this tuning in process with the Divine

[2] See Hebrews 11:1.

[3] *The Power of Your Subconscious Mind,* Joseph Murphy, Prentice-Hall, Inc. Englewood Cliffs, New Jersey, 1963.

[4] *The Miracle of Mind Dynamics,* Joseph Murphy, Prentice-Hall, Inc., Englewood Cliffs, New Jersey, 1964.

The Amazing Laws of Cosmic Mind Power, Joseph Murphy, Parker Publishing Company, Inc., West Nyack, New York, 1965.

> Power for many moons now, and I can truly say it has greatly increased my capacity for service. There is harmony in my home and office, and my employees are happier and more successful in all ways. I have greater confidence, more conviction, and greater happiness. I am closer to God in every way. It is a wonderful way of life! Kind regards, J.W., Los Angeles, California.

The writer of the above letter is a modern Los Angeles businessman. He has learned that the way to be efficient and effective in his business life is to tune in on the Cosmic Power within him. He has begun to see the order, harmony, beauty, charm, and fascination of tuning in on the indescribable wonders of this Infinite Storehouse of riches.

The Ideal Way to Tune In

Make it a regular procedure to affirm the following marvelous truths night and morning. Feel the reality of what you affirm. Charge the words with life and love, and make them meaningful in your life.

> *I know the answer to my problem lies in the Cosmic Power within me. I now get quiet, still, and relaxed. I am at peace. I know that God, the Cosmic Power, speaks in peace and not in confusion. I am now in tune with this Cosmic Power; I know and believe implicitly that It is revealing to me the perfect answer. I think about the solution to my problem. I now live in the mood I would have were my problem solved. I truly live in this abiding faith and trust, which is the mood of the solution; this is the Cosmic Power moving within me. It is omnipotent; It is manifesting in my life. My whole being rejoices in the solution; I am glad. I live in this feeling and give thanks.*
>
> *I know that God has the answer. With God, all things are possible. God is the Cosmic Power within me; It is the source of all wisdom and illumination.*

The indicator of the presence of God within me is a sense of peace and poise. I now cease all sense of strain and struggle as I tune in with this Infinite Power. I know that all the wisdom and power I need to live a glorious and successful life are within me. I relax my entire body. My faith is in this Cosmic Power; I go free. I claim and feel the peace of God flooding my mind, heart, and whole being. I know the quiet mind gets its problems solved. I now turn the request over to the Cosmic Power within me, knowing It has the answer. I am at peace.

SUMMARY

Steps in Tuning in the Cosmic Power

1. The Power that moves the world is within you. It is Almighty. Tune in on this Power, and you will release Its wonders and glories.
2. The Cosmic Power can inspire you, heal you, reveal new ideas to you, and set you on the high road to happiness, freedom, and peace of mind.
3. The Cosmic Power created you and knows all the processes of healing. One woman healed her arm by claiming and knowing, "Through the Power of the Almighty within me, I am raising my arm freely."
4. You have inexhaustible reserves within you. A general in the army depends on his reserves. You can call on your inner reserves at any time—power, wisdom, strength, guidance, freedom—they are always instantly available.
5. You can call on your reservoir of harmony, health, and

peace by using this formula: *God is, and His Presence flows through me as harmony, joy, peace, beauty, and right action. His love fills my soul, and wonders happen as I pray.*

6. When going on a journey by bus, airplane, automobile, or train, or by whatever means of conveyance, you can affirm feelingly and knowingly: *Divine love goes before me, making straight, beautiful, joyous, and happy my way.* You will bear a charmed life, and wonders will happen on your journey.
7. Prior to a school examination, claim that the infinite intelligence of your subconscious mind reveals the answers and guides you in your studies, and you will be guided to study the right material and answers will be given you—maybe in your dreams.
8. If you would like to meet a long-lost brother or sister, pray as follows: *Divine guidance is mine now, and Infinite Intelligence brings us together.* You will be amazed at how quickly the deeper currents of your mind will bring both of you together!
9. As you realize that there is no death—only life—and as you radiate love, peace, and joy to a loved one in the next dimension, rejoicing and blessing his new birthday in God, you will find that God wipes away all tears from your eyes and there is no more crying.
10. Your loved ones who have passed on to the next dimension of life are all around you, separated by a certain power frequency only. The room where you are now sitting is full of television pictures; the air outside and inside is replete with broadcasts of all kinds. Symphonies, songs, speeches, and people fill the air. If you were clairvoyant and clairaudient, you could see and hear the voices of the living and of the so-called dead.
11. Faith is a state of mind, a way of thinking, an inner certitude based on the knowledge of the laws of your

mind. The law of your Cosmic subconscious, properly used, cannot fail you.

12. Tune in on the Cosmic Power morning and night, and you will become far more efficient and successful in every way. You will become closer to God, the Giver of all Good, and wonders will happen as you pray.
13. The serene mind solves its problems. Turn over your request to the Cosmic Power, knowing you will receive the answer.

2

How your Cosmic Subconscious can guide you

All the water in the reservoir is waiting for you to turn on the tap. Turn the tap, and thousands of gallons of water are at your service. In a similar manner, all the powers of the Cosmic Kingdom within you are waiting for you to release them, and as you do the Infinite Power within you will respond and be active in your life, enabling you to bless mankind and yourself in countless ways.

Your Subconscious Can Guide You Without Error

The fields of science, art, industry, and religion need men and women of high vision, courage, confidence, and fortitude who are capable of drawing from the Infinite Powerhouse within each person.

Had Sir Winston Churchill yielded at the first assault of the enemy, England might have been lost to the invaders. However,

Churchill's insuperable faith lifted up all his countrymen and gave them a new spiritual vision.

The biographies of Abraham Lincoln point out that he was a complete failure until he went into the service of his country. You could look upon his failings as stepping stones to his success in the preservation of our Constitution. Lincoln realized and thoroughly understood that to perceive what is the right course of action and then not to follow it is the hallmark of cowardice and timorousness.

The Source of Gandhi's Tremendous Spiritual Power

Mahatma Gandhi's early life was ignominious and hellish. He was beaten, kicked, and thrown into jail, and many assassins tried to kill him before his tragic end. Unswervingly, however, Gandhi kept on with his great work, persevering and never compromising with his principles until the "untouchables" were freed from centuries of serfdom. His faith in Cosmic justice, his singleness of vision, and his courage to act brought wonderful results. Gandhi said, "Faith is nothing but a living, wide-awake awareness of God within. He who has achieved that faith wants for nothing."

How a Diplomat's Wife Increased Her Graciousness

A young lady approached me after a lecture in Town Hall, New York, and said, "A young man in the diplomatic service has asked me to marry him. I'm terrified, though, because I know I will have to entertain; but I know nothing about diplomatic etiquette or the proprieties and dictates of diplomatic society."

I suggested that she attend one of the better charm schools in New York. I had noted that her voice had a ring of acidity, and I advised her to practice continuously the imparting of loving-kindness in her words by following the injunctions of the Bible: *Let your words be like apples of gold in pictures of silver*, and *Let your words be as a honeycomb, sweet to the ear and pleasant to the bones.* (See Proverbs 25:11 and 16:24.)

Following my advice, she practiced morning, noon, and night

for several months pouring love, kindness, and goodwill into all her speech, and after a few months she noticed a tremendous change taking place for the better in her relationships with other people. She wrote me saying that the charm school had benefited her immensely. She had learned to practice all the amenities and graces of life; moreover, they taught her how to dress properly, how to walk, talk, and entertain, and how to adopt the right posture and decorum for all occasions.

I had told her that when she said "Good morning" to someone she should realize the ancient implication of the words, which mean in substance, "The Light of God shines in you," and when she said "Good night" she should feel and know that she was saying, "God giveth you sleep."

The realization of these simple truths was the turning point of her life! She is now happily married to the young diplomat and, on behalf of her country, is entertaining embassy guests with grace, charm, and dignity, and she is beloved by all.

The small everyday courtesies sweeten life; the greater ones ennoble it. Remember, too, that there is a courtesy of the heart: it is called love. From it springs the purest courtesy in outward behavior.

Hidden Talents Found Through Infinite Intelligence

A young newspaperman told me that he had been fired from his job and informed that he was a flop and a failure as a reporter. He was bitter and vindictive for a few days; then he decided to forgive his former employer and be grateful that he had been discharged. He said to himself, "I failed in that job. I didn't belong there; my talents lie elsewhere. I will be a tremendous success when I am doing what God has fitted me for."

I suggested to him that he pray frequently as follows:

> *Infinite Intelligence reveals to me my true place in life where I express my hidden talents in a wonderful way, and I follow the unmistakable lead which comes to me.*

balance. Align yourself with the Infinite Source of life and let life flow through you as abundance, security, right action, and true expression.

Thinking Makes It So—for You

Today, you are what your thoughts have made you. You are the sum total of your thinking. You have charge of your own life, and other people, conditions, and events have nothing whatever to do with your success, happiness, or destiny.

You can make yourself sick, poor, and unhappy by the nature of your habitual thinking; you can even kill yourself by your thoughts.

Affirm frequently as follows: *My faith is in God, His Cosmic Wisdom, and in all things good. I live in the joyous expectancy of the best, and marvelous and unforeseen wonders come into my life every day.* Wonders will happen in your life as you write these truths in your book of life (*your subconscious mind*). Have faith in the responses of your deeper mind to your conscious mind's thinking, and you will prosper along all lines.

How to Gain Faith Yielding Real Benefits

A man said to me, "I had absolute faith that my thoroughbred would win the race." I explained to him that it is impossible to have absolute faith in anything but the principles of life and the working of the laws of his subconscious mind. Principles and laws never change—they are eternal, immutable, and timeless. The horse in question dropped dead in the race; it is impossible to have absolute faith in the outcome of a race, an interview, or any other event in life.

I pointed out to him that his faith should be in God, His Cosmic gifts, and in all things good, and, with that accomplished, success, prosperity, and happiness would be assured him in ways he knew not of.

I explained to him about the girl who had positive faith that she would marry a certain man. The marriage was arranged, all of us waited in the church, but the groom never showed up because he had died in a taxi on the way to the chapel. She then realized that she did not control *his* life or destiny. She looked upon the matter philosophically, however, and said, "Well, God has another plan for me, and He will send me another husband, who will be my perfect partner." And, some time later she met and subsequently married a wonderful young man!

People say to me, "I had perfect faith that I would win the Irish Sweepstakes." This is not real faith, as there is no guarantee that the horse will run as you expect.

Have perfect faith in God, His love, His laws, and His Cosmic guidance, and you will never fail or want for any good thing. Have a deep-seated conviction in the laws of mind and in the ways of God, and all your ways will be paths of pleasantness and lanes of happiness.

How Faith in the God-Presence Overcame Failures

A motion picture director told me how he had prayed for success, peace, harmony, and a wonderful outcome for the picture he was directing. However, things did not turn out the way he expected! Many members of the cast got sick, and the weather on location remained very bad—the whole thing was a failure. He said to me, "I imagined and envisioned success and a constructive outcome, but I don't control the weather, the sun, moon, or stars, or the lives of the actors and actresses. I realized, though, that it is impossible for me to fail in the long run, as I believe in the principle of success, and as I am one with the Infinite. My faith is in the God-Presence, and my confidence is in the eternal truths of life which never change."

His next picture was a smashing success, and today he is one of the world's greatest directors. His faith was in the Infinite Way of Life which cannot fail, and he knows that he was born to succeed. He was not disappointed because he had a few failures. His faith was in the right place: with God and His laws which never fail.

He said to me, "I can't have absolute faith in the weather or that Tom Jones will be alive tomorrow or that the man I am going to sign a contract with will be there to sign. But I have absolute faith that God is God, and that's enough for me!" This is the wise reasoning of a great director.

How to Become Friendly, Happy, Joyous, and Free

Emerson said, "The finite alone has wrought and suffered; the Infinite lies stretched in smiling repose." Tune in on the Infinite within you now and call upon It, and the moment you align yourself with this infinite ocean of life, love, truth, and beauty, the power of God will become active and potent in your life and you will feel a deep sense of security and inner repose.

Learn to be yourself. Throw away false pride, arrogance, dissimulation, and pretense. Honor, exalt, and praise the God-Self within you. Give your allegiance, devotion, and loyalty to the God-Presence within you and recognize It as First Cause. This is loving God, and when you love God—which is the Self of you—you will be natural, real, genuine, and human, and you will enter into the joy of the Lord and the laughter of God.

SUMMARY

Powers for Winning Your Way

1. Your so-called failures are but stepping stones to your success.
2. Faith is nothing but a living, wide-awake awareness of Cosmic Power within you. He who has achieved that faith wants for nothing.
3. Practice morning, noon, and night the pouring out of

love, kindness, and goodwill to those around you, and you will notice a tremendous change taking place in your relationship with people.

4. When you seek guidance, affirm feelingly: *I am Divinely guided in all my ways. God reveals the way.* The answer will come clearly into your conscious mind.
5. You get ahead in all endeavor by adhering to the fundamental truths of life which are the same yesterday, today, and forever.
6. Focus your mind on the vision of victory and triumph, and you are sure of success. One success wipes out hundreds of failures.
7. You are what you think all day long. Align yourself with the Infinite Source of life and let life flow through you as abundance, security, right action, and true expression.
8. Affirm frequently: *My faith is in God and all things good. I live in the joyous expectancy of the best, and marvelous and unforeseen wonders come into my life every day.* As you do this, you will prosper along all lines.
9. Real faith is not in a horse, an institution, an individual, or a creed, but rather in God and the laws of your subconscious mind which never change.
10. Let your confidence be in the eternal truths which never change.

3

How to become aware of your Cosmic Power

One of the most profound and deep-seated longings of the heart is to gain a recognition of your true worth—to be respected, loved, and esteemed. Carlyle said, "One of the Godlike things of this world is the veneration done to human worth by the hearts of men."

The Psalmist reminds man to appreciate his true self with these majestic words of wisdom in Psalm 8:3-8:

When I consider thy heavens, the work of thy fingers, the moon and the stars, which thou hast ordained; What is man, that thou art mindful of him? and the son of man, that thou visitest him? For thou hast made him a little lower than the angels, and hast crowned him with glory and honor. Thou madest him to have dominion over the works of thy hands; thou hast put all things under his feet: All sheep and oxen, yea, and the beasts of the field; The fowl of the air, and the fish of the sea, and whatsoever passeth through the paths of the seas.

Here David speaks eloquently and beautifully of the tremendous potentialities within man, and today we see man exploring space, so successfully that in our time he will undoubtedly visit the moon and other planets. Today we are witnessing the workings of Infinite Intelligence through man in countless new discoveries. Scientists inform us that we are living in an age of light and supersonic speed, electronics and electricity, radio and radar—all of these miracles of the air, space, and sea come out of the mind of man.

Recently, a mathematician said to me that only abstractions can explain the world and that events which transpire today on land, on the sea, and in the air can be understood only by physicists and mathematicians.

The Powers Within

Today man is also penetrating and navigating the waters of his deeper mind and is gradually becoming aware of the Kingdom of God within him. Research work at Duke University and other academic laboratories is revealing the powers of man's mind in such discoveries as telepathy, clairvoyance, clairaudience, telekinesis, extrasensory travel, precognition, retrocognition, and other marvelous powers of the mind.

Self-Esteem Gained Through Cosmic Awareness

Recently I received a letter from a woman in Arizona in which she said that her sister-in-law and mother-in-law disapproved of her and had told her bluntly that they preferred her husband's former wife. They never invited her to their homes but always asked her husband to visit them alone. Furthermore, although she tried her utmost to be nice to them, they criticized her meals, her home, her clothing and her speech. This woman said that she felt inferior and rejected and asked me, "Why do they do this? What's wrong with me?"

In reply I pointed out that she had been suffering unnecessarily and without real warrant, and that she had the power to refuse and to reject the poisonous statements of her in-laws, their rudeness, and their incivility. I explained further to her that she did not create her mother-in-law or sister-in-law and that she was not responsible for their jealous, envious attitudes and complexes. I told her to stop placing them on a pedestal and to stop being a doormat for them—that a doormat is something one walks on.

I added that it was quite possible that her charm, graciousness, kindness, and wonderful character annoyed them and that they probably got a sadistic satisfaction in disturbing her.

I suggested that she break off all relations with them and cease abasing herself by toadying to them. I told her that she needed an attitude of self-respect and self-esteem. I gave her the following prayer to be used three times a day:

> *I completely surrender my in-laws to God. God made them and He sustains them. I radiate love, peace, and goodwill to them, and I wish for them all the blessings of Heaven. I am a child of God. God loves me and cares for me. When a negative thought of anger, fear, self-criticism, self-condemnation, or resentment enters my mind, I immediately supplant it with the thought of God in the midst of me. I know I have complete dominion over my thoughts and emotions. I am a channel of the Divine. I now redirect all my feelings and emotions along harmonious, constructive lines. Only God's ideas enter my mind, bringing me harmony, health, and peace. Whenever I am prone to demean or to demote myself I will boldly affirm: "I exalt God in the midst of me. I am one with God, and one with God is a majority. If God be for me, who can be against me?"*

She faithfully followed the above prayer and also carried out my other instructions.

A few days ago, I received her story of answered prayer:

> Dear Dr. Murphy: Thank you for your letter and the enclosed prayer. I phoned my in-laws and told them not to visit us at any time except when my husband and I issued

a special invitation. I also told them that I sincerely wished them well—and I meant it. I see now where I made my mistake, and how I was actually handicapping myself by thinking I was inferior to them.

The prayer has worked wonders for me, and my husband said to me the other day, "Honey, you are beaming. What happened to you?" I told him. Both of us are so grateful. Signed, Mrs. L.M.

Shyness and Timidity Overcome by Using the Cosmic Power

Some time ago, I had a consultation with a salesman who said that he was timid, shy, and resentful, and looked upon the world as harsh and cruel. Actually, he was trying to escape from taking his rightful dominion over life: he said that his wife, his boss, and his associates did not appreciate him and that his children looked down on him.

[illegible]e [illegible]ause of all this was that this young man had an inner sense [illegible] insecurity and inadequacy, and he was down on himself. He ask[illegible] How can I gain the appreciation of others?"

I found myself reminding him of this great Scripture passage: *Love thy neighbor as thyself* (Mark 12:33). The real meaning of this text is that your neighbor *is* yourself, because the real Self of you is God. Tennyson said: Speak thou to him for He hears, and Spirit and Spirit can meet/Closer is He than breathing, and nearer than hands and feet.[1]

Another everyday meaning of the text is to love your neighbor as you love yourself. I explained to this young man the truth about himself and how to love and appreciate himself more along the following lines: If a man demotes, despises, and deprecates himself, he can't lift up or give esteem, goodwill, and respect to others, *for it is a cosmic law of mind that man is constantly projecting his thoughts, feelings, and beliefs onto others and what he sends out comes back to him.*

Man is a son of the Infinite, and all the qualities and powers

[1] "The Higher Pantheum," Stanza 6.

of God are within man, waiting to be expressed. Man must love and honor the indwelling God.

The True Meaning of Self-Love

Love of self in the true Biblical meaning is to honor, recognize, exalt, respect, and give your total allegiance to the living Spirit within you. This supreme Intelligence made you, created you, animates, and sustains you. It is the Life-Principle within you. This has nothing to do with egoism or self-aggrandizement, but on the contrary is a wholesome veneration for the Divinity which shapes your ends. The Bible says that your body is a temple of God, and therefore, as Paul says, you are to glorify God in your body, also (*see* I Corinthians 6:20). When you honor, respect, and love the Self of you, you will automatically love, esteem, and honor others.

This salesman listened carefully and avidly and then said to me, "I never heard it explained that way before. I can see clearly what I have been doing. I have been down on myself, and I have been full of prejudices, ill will, and bitterness, and what I have been sending out has reverberated back to me. I have gained a true insight into myself."

This salesman practiced affirming the following truths with deep sincerity several times daily, knowing that they would sink down from his conscious to his subconscious mind and like seeds come forth after their kind:

> *I know that I can give only what I have. From this moment forward, I am going to have a wholesome, reverent, and deep respect for my real Self, which is God. I am an expression of God and God hath need of me where I am, otherwise I would not be here. From this moment forward I honor, respect, and salute the Divinity in all my associates and all people everywhere. I hold the Self of every person in veneration and esteem. I am one with the Infinite. I am a tremendous success,*

and I wish for all men what I wish for myself. I am at peace.

Learn to Love Yourself

The above young man has transformed his life; he is no longer timid, shy, or resentful. He has gone ahead by leaps and bounds—and so can you! Learn to love your true Self, and then you will learn to love and respect others.

What thou seest, man,
That too become thou must:
God, if thou seest God,
Dust, if thou seest dust.

—*Anonymous*

Self-Condemnation and Annoyance Overcome

Some months ago I had a letter from a man who stated that he couldn't understand why everybody around him annoyed him. I asked him to come and see me and in talking with him I discovered that he was constantly rubbing others the wrong way. He did not like himself and was full of self-condemnation. He spoke in a very tense, irritable tone. His acerbity of speech grated on one's nerves. He thought meanly of himself and was highly critical of others.

I explained to him that, while his unhappy experiences seemed to be with other people, his relationship with them was determined by his thoughts and feelings about himself and them. I elaborated on the fact that if he despises himself he cannot have goodwill and respect for others, because it is a law of mind that he is always projecting his thoughts and feelings on to his associates and all those around him.

He began to realize that as long as he projected feelings of prejudice, ill will, and contempt for others, that is exactly what he would get back because his world is but an echo of his moods and attitudes.

Practicing the Golden Rule

I gave him a mental and spiritual formula which enabled him to overcome his irritation and arrogance. He decided to write consciously the following thoughts in his subconscious mind:

> *I practice the Golden Rule from now on, which means that I think, speak, and act toward others as I wish others to think, speak, and act toward me. I walk serenely on my way, and I am free, for I give freedom to all. I sincerely wish peace, prosperity, and success to all. I am always poised, serene, and calm. The peace of God floods my mind and my whole being.*
>
> *Others appreciate and respect me as I appreciate myself. Life is honoring me greatly, for it has provided for me abundantly. The petty things of life no longer irritate me. When fear, worry, doubt, or criticism by others knock at my door, faith in goodness, truth, and beauty opens the door of my mind, and there is no one there. The suggestions and statements of others have no power. The only power is in my own thought. When I think God's thoughts, God's power is with my thoughts of good.*

He affirmed these truths morning, noon, and night, and he committed the whole prayer to memory. He poured into these words life, love, and meaning, and by osmosis these ideas gradually penetrated the layers of his subconscious mind. His letter speaks for itself:

> Dear Dr. Murphy: May I first say thank you for all the calm and happy feelings I now possess. I know full well that this is brought about by my new understanding of my mind and how it works. I know why I have a high opinion of my-

self and of all human beings. I honor myself, and in so doing I know I am honoring God.

I am learning how to specialize myself out of the law of averages. I am getting along fine and have received two promotions in the past two months! I now know the truth of this passage: *I, if I be lifted up . . . will draw all men unto me.*[2] Gratefully, E.J.

The above letter demonstrates how any man can overcome irritation and annoyance. This young man learned that the trouble was within himself, and he decided to change his thoughts, feelings, and reactions. Any man can do the same. It takes decision, stick-to-it-iveness, and a keen desire to transform oneself. *Go, and do thou likewise* (Luke 10:37).

Look at the Small End

An astronomer friend of mine said to me that for years he had scanned the heavens, seeking the answer to the story of creation and the riddle of the universe through the telescope, but that lately he has been looking within himself, who is inevitably at the small end of the telescope. He added that the small end of the telescope is the important end, for within man is God, and the entire secret of creation and the mystery of the Cosmos.

When man learns about himself, he will have learned about the universe. Now it is time to analyze the analyzer. In trying to find happiness, peace, and prosperity outside of himself, man has neglected to look within himself to the infinite storehouse of riches within his subconscious mind.[3]

Where will you find poise, balance, peace, and happiness but in your own mind, through your thoughts, feelings, and a sense of oneness with the eternal verities and spiritual values of life? William Shakespeare said: "What a piece of work is a man! How noble in reason; how infinite in faculty! In form and moving, how

[2] John 12:32.
[3] *The Power of Your Subconscious Mind,* Joseph Murphy, Prentice-Hall, Inc., Englewood Cliffs, New Jersey, 1963.

express and admirable! In action how like an angel; in apprehension how like a god." [4]

How to Get a Higher Estimate of Yourself

Ralph Waldo Emerson said: "There is one mind common to all individual men, and every man is an inlet to the same and all of the same." He also said, "He who is admitted to the right of reason is a freeman of the whole estate."

Begin to believe this. Realize that Infinite Intelligence—the guiding principle of the universe—is within you, and the infinite healing Presence of God controls all your vital organs and all the processes and functions of your body. You have the capacity to make choices, to use your imagination and all the other powers of God within you. Your mind is actually God's mind. When you consciously, decisively, and constructively use the Infinite Wisdom within, you become *a free man of the whole estate.*

Emerson inspires you to enlarge the concept of yourself when he announces this profound truth: "What Plato has thought, man may think; what a saint has felt, he may feel; what at any time has fallen any man, he can understand. Who hath access to the Universal Mind is a party to all that is, or can be done, for this is the only and sovereign agent."

Emerson was America's greatest philosopher and one of the greatest thinkers of all time. He was constantly in tune with the Infinite, and he urged all men to release the infinite possibilities within them. Emerson taught the dignity and grandeur of man and pointed out to his listeners that the great appear great to us only because we are on our knees—that we attribute greatness to Plato and others because they acted upon what they themselves thought, and not upon what other people believed or what others thought they should think.

Begin to have a lofty, noble, and dignified concept of yourself, and remember what the Psalmist said to all men: *I have said, Ye are gods; and all of you are children of the most High.* (Psalm 82:6)

[4] *Hamlet,* II, ii.

Health Improved with New Self-Appraisal

The following letter speaks for itself:

> Dear Dr. Murphy: This is to thank you for writing *The Miracle of Mind Dynamics.*[5] I have read it over and over about sixteen times, but more than that I have applied it. I have stopped whining and complaining, and I am no longer bitter, angry, or hateful.
>
> My husband left me a year ago for a younger woman. I suffered from such intense rage that my doctor said the sudden precipitation of arthritis was caused by my emotional shock, anger, and hatred. Every day for the past three months I have claimed boldly as you suggest that my body is a temple of the living God and that I glorify God in my body. Every day for the past several months, for about fifteen minutes every morning, afternoon, and evening, I affirmed that God's love permeated every atom of my being and His Heavenly Presence saturated my whole being. I also prayed for my ex-husband.
>
> There has been a remarkable change in my body; the edema and excruciating pain have subsided, the suppleness and mobility of my joints have improved remarkably, and the calcareous deposits are gradually disappearing. My doctor is delighted and so am I.
>
> I continue to realize that I am a child of God and that God loves me and cares for me. I know that this new self-appraisal has wrought wonders in my life. All hatred of my ex-husband has gone, and I am on the way to perfect health. Divine law and order govern me.
>
> I am eternally grateful for your writings. Mrs. W.M.

This woman discovered what the power of a true appraisal of her real Self can do. She found that when she began to think of herself as the temple wherein God dwells, and as she began to honor, exalt, and call upon this Divine Presence, It responded as the emotions of love, peace, confidence, joy, vitality, wholeness, and goodwill.

[5] *The Miracle of Mind Dynamics,* Joseph Murphy, Prentice-Hall, Inc., Englewood Cliffs, New Jersey, 1964.

As she began to love and respect herself, she discovered that all hatred vanished and that love rushed in to fill up the vacuum. Love is the fulfilling of the law of health, happiness, success, and prosperity.

Formula for Business Success

A prominent businessman in Los Angeles told me that the secret of his success and prosperity is that he learned a great truth which he demonstrates daily. This is his formula:

> *I know that the God in the other is the same God in me, and therefore if I hurt the other I would be hurting myself and that would be foolish. Knowing this I practice the greatest formula of all. I bless and exalt the good in the other. I make it a point to advance his interests and I know as I do I am advancing my own. I know that the ship which comes home to my brother comes home to me.*

Practice the above formula and you will like yourself more. You will see sermons in stones, tongues in trees, songs in running brooks, and God in everything, including your fellow man.

SUMMARY

Profitable Pointers in This Chapter

1. One of the most deep-seated longings of the heart is to gain recognition of one's true worth—to be respected, esteemed, and loved.

2. Academic laboratories are revealing the tremendous powers within man, such as his ability to see, hear, feel, and travel independent of the physical senses and organs.
3. You have the power to mentally reject and neutralize all negative suggestions or critical remarks of others. Whenever you are prone to criticize or condemn yourself, affirm: *I exalt God in the midst of me.*
4. *"Love thy neighbor as thyself"* means that you honor, exalt, appreciate, love, and give complete allegiance to the God-Power within you—your real Self.
5. True self-love has nothing to do with egotism, self-aggrandizement, or morbid selfishness. On the contrary, it is a wholesome veneration for the indwelling God which is the Reality of all men.
6. What you send out reverberates back to you. Life is an echo; therefore, send forth love, peace, goodwill, and blessings to all men and to the whole world, and countless blessings will return to you.
7. Develop self-esteem by dwelling consciously on the following truths: *I know that I can give only what I have. From this moment forward, I am going to have a wholesome, reverent, and deep respect for my real self, which is God. I hold the Self of every man in veneration and respect.*
8. What thou seest, man,/That too, become thou must;/ God, if thou seest God,/Dust if thou seest dust.
9. When you think meanly of yourself, you can't think well of others inasmuch as you always project to others your own thoughts, feelings, and beliefs.
10. You practice the Golden Rule as you begin to think, speak, and act toward others as you wish others to think, speak, and act toward you. Begin to wish for all men what you wish for yourself, and countless blessings will be yours.

11. God indwells all men. The Kingdom of God is within. The entire secret of creation and the mystery of the Cosmos are within man. When man learns about himself, he will have learned about the universe. Now is the time to analyze the analyzer.
12. There is one Mind common to all individual men, and every man is the inlet and outlet to the same.
13. Begin to have a lofty, noble, dignified concept of yourself. Remember the great truth of the Bible: *I have said, Ye are gods; and all of you are children of the most High.* (Psalm 82:6)
14. As you begin to love, respect, and exalt God in the midst of you, all bitterness and hatred will be dissipated. Love is the fulfilling of the law of health, happiness, and peace of mind.
15. The greatest formula for self-appreciation and respect for others is to realize that the God in you is the same God in the other, and that as you advance the interests of another you are advancing your own. The ship that comes home to your brother comes home to you.

4

How the Cosmic Power can solve problems

Ninety Per Cent of Problems Are Humanly Created

Change your thoughts and keep them changed, and you can transform your whole life! Mr. Fred Reinecke, an engineer, said to me recently that ninety per cent of the problems of the men in a plant were due to their personality defects and to their inability to get along with others. Then he added that only about ten per cent of the problems were of a technical nature.

The Right Way to Solve Problems

There is a right way to talk, to walk, to drive a car, to bake a cake. Actually, there is a right and wrong way to do everything.

To live a full and happy life you must live according to certain immutable and eternal principles. You would not think of building a wheel off-center or of violating the principles of electricity or chemistry. Likewise, when you think, speak, act, and react from the standpoint of the Infinite Intelligence within you, you will find that your whole life will be one of joy, happiness, success, and peace of mind.

How Changed Thinking Healed Ulcers and High Blood Pressure

Mrs. Wrongway was jealous and hateful toward the supervisor in her office; she had developed ulcers and high blood pressure. However, she became interested in the spiritual principle of forgiveness and goodwill. She realized that she had accumulated many resentful and grudging attitudes and that these negative and obnoxious thoughts were festering in her subconscious mind. She tried to talk with her supervisor in an effort to straighten matters out, but the woman brushed her off.

In a continuing effort to correct the situation, Mrs. Wrongway applied the principles of harmony and goodwill for ten minutes every morning prior to going to work. She affirmed as follows: *I surround Mrs. X with harmony, love, peace, joy and goodwill. There are harmony, peace, and understanding between us, and whenever I think of Mrs. X I will say, "God's love saturates your mind."*

A few weeks passed, and Mrs. Wrongway went to San Francisco for a weekend. On boarding the plane, she discovered that the only vacant seat was the one next to her supervisor! She greeted her cordially and received a cordial and loving response. They had a harmonious and joyous time together in San Francisco. They are now attending my lectures together. Infinite Intelligence set the stage for the solution of this difficulty. Mrs. Wrongway's changed thinking had changed everything, including a perfect healing of her ulcers and high blood pressure.

How the Cosmic Power Effected a Promotion

I recall a young lady once telling me, in my study: "Everybody in my office dislikes me; there are several who even want me fired."

I said to her, "Why don't you resign and find another position?"

"What's the use?" she said. "I've had six jobs this year so far."

This young lady had a brilliant mind, was well educated, and was an outstanding legal secretary. Ninety per cent of her problem was in her personality.

I gave her a spiritual prescription and suggested that she take it regularly at least morning and night for several months. I told her to pray the following prayer for every man and woman in her office every day before she went to work:

> *I send out loving thoughts and feeling of goodwill, happiness, and joy to all those in my office. I affirm, claim, and believe that my relationship with each one of my co-workers will be harmonious, pleasant, and satisfactory. Divine love, harmony, peace, and beauty flow through my thoughts, words, and deeds, and I am constantly releasing the imprisoned splendor within me. I am happy, joyous, and free, bubbling over with enthusiasm, and I rejoice in the goodness of God in the land of the living and in the innate goodness of all people.*

At the end of two months she received a wonderful promotion and was put in complete charge of the entire legal office.

How a Mother's Mental Movie Worked Wonders

A mother in Beverly Hills was worried about her son failing in his final examination in medical college. At my suggestion, she stopped nagging and worrying about the boy, and instead began to paint a mental picture of graduation day when her son would receive his diploma.

Several times daily, she visualized the scene clearly, feeling the naturalness and joy of the whole procedure. She congratulated her son in her imagination and made a habit of this by practicing her mental movie for at least ten minutes three times daily. Whenever she was prone to worry, she flashed on the movie in her mind. She constantly saw the accomplished fact. She imagined being present at his graduation, and she saw the fulfillment of her dream.

Incidentally, I might add that shortly after her change of mental attitude, her son became intensely interested in his studies and showed remarkable improvement along all lines. Her feeling of success and triumph was subjectively communicated with Cosmic Power to her son, and he responded accordingly.

A Healing of Personality for Business Success

A man whom I was interviewing recently said to me, "I'm all mixed up and tied up. I can't get along with others; I'm constantly rubbing them the wrong way."

This young man was hypersensitive, jittery, self-centered, and crotchety. In spite of all this, he wanted to have good relations with his co-workers and to get along well with them in every respect.

I explained to him that his present personality represented the sum total of his habitual thinking, early training, indoctrination, and emotional atmosphere plus the sum total of beliefs inculcated upon his mind but that he could transform himself. I explained to him that God indwells him and that all the attributes, potencies, qualities, and aspects of God were lodged in his deeper mind and could be resurrected and expressed in his personal life.

Accordingly, I gave him the following specific prayer for the purpose of transforming his entire personality and making him radiant, happy, joyous, and fabulously successful.

He affirmed feelingly and lovingly several times daily as follows:

> *God is the Great Personality, the One Life being expressed through me. God is, and His Presence flows through me now as harmony, joy, peace, love, beauty, and power, and I am a channel for the Divine. His wholeness, beauty, and perfection are constantly being expressed through me. Today I am reborn spiritually! I completely detach myself from the old way of thinking, and I bring Divine love, light, and truth definitely*

into my experience. I consciously feel love for everyone I meet. Mentally I say to everyone I contact, "I see the God in you, and I know you see the God in me." I recognize the qualities of God in everyone. I practice this morning, noon, and night; it is a living part of me.

I am reborn spiritually now, because all day long I practice the Presence of God. No matter what I am doing—whether I am walking along the street, shopping, or going about my daily business—whenever my thought wanders away from God or the Good, I bring it back to the contemplation of His Holy Presence. I feel noble, dignified, and God-like. I walk in a high mood, sensing my oneness with God. His peace fills my soul.

As this man made a habit of allowing attributes and qualities of Cosmic Good to flow through his mind, his whole personality underwent a marvelous change. He became affable, amiable, urbane, and increased in understanding, and he now communicates vibrancy and goodwill wherever he goes. In addition, he has moved up several rungs on the ladder of success in his field of work.

How You Can Develop a Marvelous Personality

Emerson said that religion includes the personality of God. All the elements of personality are within God, such as love, peace, joy, beauty, laughter, happiness, power, harmony, rhythm, order, serenity, and proportion. God is also Law, and we express our Divine personality as we begin to operate the law and claim, feel, and know that these qualities and attributes are flowing through us. Then we become more God-like every day.

The word "person" comes from the Latin *persona*, which means a mask originally worn by Greek actors. In ancient times the Greek actor put on a mask and assumed the role of the person depicted by the mask. He dramatized through the mask the characteristics and qualities of the personality it suggested.

How a Secretary Practiced Empathy and Established Harmony and Understanding

The dictionary defines empathy as "the imaginative projection of one's own consciousness into another being." It might be called "sympathetic understanding."

Mrs. Jean Wright, secretary of our organization, told me how she practiced this art in relation to another girl in an office where she formerly worked. This girl was very hostile, antagonistic, and quite obstreperous; a great misunderstanding seemed to be growing between them.

As a result, Mrs. Wright sat still a few times each day and projected herself into the mind of the other woman and looked out at herself through the other's eyes; she then corrected what she saw and claimed: *"God's peace, harmony, and understanding reign supreme between us. Whenever I think of Miss S., I claim, 'She is loving, kind, cooperative, and harmonious.'"*

After about a week, the girl invited Mrs. Wright to her home for dinner; during the visit they were pleasantly surprised to discover that they had many interests in common. Eventually, these two women became fast friends as well as good associates in their work.

How a Silence Between Sisters was Broken

Recently, while I was speaking in San Francisco under the auspices of The Institute of Religious Science, an old friend had breakfast with me at my hotel. She told me that her only sister would no longer speak to her; when she phoned her sister, the latter would hang up the phone with a curt phrase: "I'm busy; don't bother me."

My friend found it very difficult to understand her sister's attitude, which on the surface seemed so unreasonable and stupid. I spoke to her, pointing out that if her sister had tuberculosis or cancer, she would not be angry with her. My friend answered, "Of course not. I would be most compassionate."

Then I added, "Your sister has what we might call tuberculosis of the mind, and you must realize that many people have twisted, morbid, and distorted mentalities and are often referred to as mental hunchbacks."

Suddenly she realized that she was not responsible for the mental state of her sister, no more than she might be responsible for someone who was an alcoholic, a schizophrenic, a psychotic, or a paranoid.

She said, "Oh, I see now! I am not responsible for her mental sickness and hostility toward me, and all I owe her, as Paul said, is love. I do love her and wish for her all the blessings of life."

She prayed as follows: *I completely surrender my sister to God. I radiate love, peace, and goodwill to her, and there are harmony, peace, and Divine understanding between us. I loose her and let her go.*

After a few days, her sister called and apologized for her rudeness and hostility. She freely admitted that the break-up of her engagement had caused her to project her resentment and hostility not only to her sister but also to many others.

I might add that my friend was spiritually and emotionally mature, but she was overly concerned with the deformed and warped mentality of her sister.

Your understanding will keep you from anger, criticism, and hatred, or from seeking retaliation against the warped minds of others. Remember, you would not hate a person because of a physical deformity, such as that endured by a hunchback; rather you would be grateful that you were spared from such a misfortune.

The person with a distorted personality is very unhappy within himself and is a seething cauldron of inner turmoil. Quite frequently, he lashes out at those who have been most kind and generous to him, because their inner serenity, tranquility, and poise reveal his own disturbed emotional state; since he can't reach their quietude, he unconsciously tries to drag them down to his own emotional debauch. Misery loves company!

His Application of the Golden Rule Resulted in a Wonderful Promotion

I talked some months ago with a man who had earned a Ph.D. and had believed that his diploma would guarantee his success, prestige, and recognition; but he eventually discovered that there were men in his organization who had no degree at all and who earned far more money than he did and exercised greater responsibility.

I explained to this young man that there are many distinguished Ph.D.'s, linguists, professors, doctors, and similar brilliant and capable men from all walks of life who now inhabit Skid Row in Los Angeles and the Bowery in New York. These men usually attribute their trouble to alcohol or women; but the real reason is self-depreciation, self-condemnation, self-loathing, and lack of all contact with the treasure and wisdom of their Cosmic Powers, which could set them on the highroad to happiness, freedom, and peace of mind. They are down on themselves and they lack self-acceptance and true expression. Their habitual thinking attracts to them misery, suffering, and poverty. Yet these men, who are often highly educated, have at one time risen to the higher echelons in their respective fields. Their alcoholism, dope addiction, and abnormal behavior are but symptoms of their twisted and warped personalities.

The man of whom I am speaking was highly critical and jealous of his associates. He spoke of them with a touch of asperity in his voice. His trouble was his failure to practice the Golden Rule.

At my suggestion he began to treat others as he would wish others to treat him. He began to practice smiling and being kind, graceful, and pleasant with all his associates. He persisted in this attitude until it became habitual, and today he is one of the most affable, amiable, and loving personalities you could meet. His success is assured, and he has received a wonderful promotion which dramatizes the truth: change your thought pattern and you change your life!

You Are Needed Regardless of Your Age

Emerson said, "The Great Oversoul has need of an organ where I am, else I would not be here." Sometimes I meet people who say, "Oh, my children are all grown up. They don't want me or visit me any more." Remember, God and His Cosmic World have need of you. There is no such thing as an unneeded or an unwanted person. Each person is a note in the grand symphony of all creation. There is a special and particular role for you to play. The Bible says, *The voice of him that crieth in the wilderness, Prepare ye the way of the Lord, make straight in the desert a highway for our God* (Isaiah 40:3).

The desert of confusion, lack, and limitation is your own mind, and you can now listen to the inner voice which speaks to you in the form of your heart's desire and which is telling you to arise and to elevate yourself higher because God has need of you. Say to yourself: "God gave me this desire, and He reveals to me the perfect plan for its fulfillment in Divine order." As you adhere to this truth, the way will open up and your desert will rejoice and blossom as the rose.

Saying "Yes!" to Life in Cosmic Truth

The Bible says: *Let your communication be, Yea, yea; Nay, nay: for whatsoever is more than these cometh of evil* (Matthew 5:37).

Say "yes" to all the ideas which heal, bless, inspire, elevate, and strengthen your life. Come to a clear-cut decision that you will accept only the eternal verities and spiritual values of life and then purposefully build these into your personality. Rejoice in your sense of oneness with the Infinite Intelligence and Infinite Mind and contemplate the wonders of your Divine sonship. Say "no" boldly to all teaching, ideas, thoughts, creeds, and dogmas which inhibit, restrict, and instill fear into your mind. In other words, accept nothing mentally that does not fill your soul with joy.

Realize that God is Life, and that He is your life now. God is Love, and His love fills your soul. God is Joy, and you are expressing the fulness of joy. God is Wisdom, and your intellect is anointed constantly with the Light from On High. God is peace, and you are expressing more and more of His peace in your thoughts, words, and deeds. As you make a daily habit of realizing these truths, you will develop a radiant personality and make a highway for your good in all ways and things.

How You Can Triumph Over Depression and All Obstacles

The Bible says: *Every valley shall be exalted, and every mountain and hill shall be made low: and the crooked shall be made straight, and the rough places plain.* (Isaiah 40:4).

When you are in the valley of despair, dejection, and melancholia, turn to the God-Presence within you and realize that external things and conditions are not causative. All things pass away. Conditions do not create other conditions.

The primary *cause* is your thought and feeling—i.e., your mental attitude, the way you believe. Conditions and circumstances are suggestive only; you have the power to reject or to accept them.

Decide, therefore, that Infinite Intelligence reveals the way out. Contemplate in your Cosmic thought the way you want things to be, and the mountain (problem) will be removed and the hill (obstacle or difficulty) will be shattered. As you claim that Divine law and order govern your life, the crooked (the ups and downs of life, the swings of fortune) shall be made straight and the rough places plain—i.e., you will begin to live a balanced life of growth, achievement, and advancement free from detours into sickness, accidents, losses, and foolish expenditures of energy, time, and effort.

As you keep your eyes on the Cosmic Power, and as you tune in on the Universal Wisdom within you and make contact through your thought and feeling, all the barriers, delays, impediments, and difficulties will disappear, and the desert of your life will truly rejoice and blossom as the rose.

SUMMARY

Your Aids to Authority

1. Change your pattern of thought-life and you will change your destiny.
2. There is a right way and a wrong way to use your mind. To live a full and happy life you must live according to principles of Cosmic Wisdom.
3. Infinite Intelligence will respond to you when you appeal to It, and set the appropriate stage for the solution of your problem.
4. A wonderful spiritual prescription is to pray for all those with whom you work, wishing for them all the blessings of life. What you wish for others, you also attract to yourself.
5. Flash a mental movie in your mind and see the happy ending. Picture frequently the fulfillment of your dream, and it will come to pass in your experience.
6. As you make a habit of allowing God's attributes and love for all to flow through your mind and heart, you will develop a new, marvelous and unique personality.
7. If you can't get along with another person, practice empathy by practicing sympathetic understanding. Look at the situation through the other's mind.
8. Don't be overly concerned with the warped mentality of a relative, because you are not responsible for it. Radiate peace, love, and goodwill, and a Divine understanding will correct the situation.
9. The Great Oversoul of Cosmic Wisdom, or God, has

need of you where you are; otherwise, you would not be here. You are wanted and needed now!

10. Say "Yes!" to all those Cosmic ideas which bless, heal, inspire, elevate, and dignify your soul.
11. When you are melancholy or depressed, turn to the Cosmic Power within you and let God's river of serenity and love fill your soul. Realize that all things pass away, and the day breaks and the shadows flee away.

5

How to use the Cosmic Healing Power

All over the world today men and women of all creeds are awakening to the tremendous therapeutic results following the application of mental and spiritual laws. In the fields of medicine, psychiatry, psychology, and other related fields, evidence is being adduced and articles written on the destructive effects of mental and emotional conflicts behind various diseases.

When the conscious mind of the sick person redirects his subconscious mind along God-like channels, a cleansing takes place, and the Infinite Healing Presence is released, causing the miracle of healing to occur.

How a Crippled Hand Was Healed

Following is a verbatim testimonial of the response of the Infinite Healing Presence when you call upon It:

> Dear Dr. Murphy: I broke my left wrist, and the bones in my wrist and hand were so shattered that the doctor had to set them under a fluoroscope. I was told my hand would be

crippled and I would have to learn to compensate with my other hand. Since I do secretarial work, I must have the use of both hands. The Healing Power mended and renewed the bones and muscles. In three and one-half months I was back at work. During that time, I had repeated feelingly many times a day, *The Creative Intelligence which made my wrist is healing me now.*

The doctor had told me I would have arthritis in my wrist and that a change of weather would cause it to ache. That was seven years ago. Today, I have full use of my hand, no arthritis, and change of weather makes no difference. In fact, my left wrist and hand are more flexible and nimble than my right hand!

Again, I thank you for your prayers and instructions on how to use the Healing Power within me. Sincerely, Mrs. M.D.B.

I might add to this lady's letter by pointing out that she has a thorough knowledge of the conscious and subconscious mind and is a deep student of *The Power of Your Subconscious Mind.*[1] She realized that spiritual therapy is the synchronized, harmonious, and intelligent function of the conscious and subconscious levels of mind, specifically directed for a definite purpose.

You will notice that she completely rejected the prognosis of a crippled wrist and affirmed with deep understanding that the Infinite Intelligence which created her hand was healing it. These ideas sank into her subconscious mind, and the healing followed.

How Cosmic Power Healed a Diseased Kidney and a Broken Bone

The following is a letter from a wonderful woman who understands the Healing Power resident in the Cosmic Mind:

Dear Dr. Murphy: You have helped me so many, many times, it is hard to select the most important occasion. I recall being in terrible pain, and while our family physician was away on vacation two strange doctors told me I had a

[1] *The Power of Your Subconscious Mind,* Joseph Murphy, Prentice-Hall, Inc., Englewood Cliffs, New Jersey, 1963.

cyst on the right ovary and an abscess of a kidney. I prayed, and the pain disappeared overnight and an examination showed the abscess and the cyst had disappeared. I claimed: *God's healing love in my subconscious mind is this moment dissolving everything unlike itself.* I felt the truth of this, and the miracle happened. I know you prayed, also, and I thank you.

The most wonderful demonstration was for my mother. When she was 85 she fell and broke her pelvic bone and shoulder blade. Pneumonia set in, and the doctor did not think she would live—and if she did, she would not walk again but would have to be in a wheel chair. She said, "That's his opinion," and through faith and your prayers she was walking again in three months—without even the use of a cane!

Of course, you may use our names. Mother's was Bertha Sparrow. Signed, Mrs. Eric B. Marlor, Los Angeles, California 90004.

The Cosmic Healing Power and How You Can Use It

The true method of spiritual healing does not lie in some magic wand-waving, but rather in the mental response of man to the indwelling Cosmic Power which created him and all things in the world.

Spiritual healing is not the same as faith healing. A faith healer may be any person who heals without any knowledge or scientific understanding of the powers of the conscious and the subconscious mind. He may claim that he has some magical gift of healing, and the sick person's blind belief in him or his powers may bring results.

The spiritual therapist must know what he is doing and why he is doing it. He trusts the law of healing. The law of mind is that whatever you impress on your subconscious mind, will be expressed in like manner as form, function, experience, and events.

Impress your subconscious with thoughts of peace, harmony, health, and perfection by dwelling sincerely on these concepts with interest. Your thought and feeling (interest) will induce a healing response from the Cosmic Power.

How the Cosmic Power Healed a Tailor's Blindness

Recently I interviewed a tailor who was gradually going blind. He had had several retinal (eye) hemorrhages, and his doctor had suggested that he give up tailoring and live in the country. He stayed with his work, and his sight was growing worse. His doctor, an old friend of mine, suggested to the tailor that he talk to me about his home life.

In discussing his eyes, the tailor admitted to me that he hated the sight of his mother-in-law, who had lived in his home for several years and was habitually petulant, crotchety, cantankerous, and a source of endless trouble. I suggested that he ask her to leave at once, which he did; at the same time, he wished for her all the blessings of life, thereby eradicating all the resentment and hostility from his subconscious mind.

For ten or fifteen minutes every day he affirmed boldly: *By day and by night I am seeing more and more of God's love, light, truth and beauty in every person and in every thing. God is healing me now, and I give thanks for my perfect vision.*

In due course I received the following note from him:

> Dear Dr. Murphy: I am writing to thank you for the healing of my eyes. You opened my eyes when you explained to me that I was the cause of my own eye trouble. My doctor said my vision is normal now, and I don't have to give up tailoring. God bless you. E.S.

Twentieth-Century Miracles of Healing

A miracle is not a violation of natural law. A miracle does not prove that which is impossible; it proves that which is possible. A miracle is something that happens when one brings in a higher law than those which man has known heretofore.

How the Cosmic Power Resolved a Writer's Manuscript Problem

The following paragraphs are taken verbatim from Mrs. Arnold's letter, showing how she used the healing principle to resolve a frustration. She had spent a great deal of time and care in writing a manuscript, but she felt the title was completely wrong.

> Dear Dr. Murphy: . . . For several weeks I have been writing down titles, scratching them out, and trying again. Sunday I had a strong compulsion to get to church, which I had to hurry to do as household duties, etc., had made me somewhat late in getting started.
>
> When I got into the Ebell Theatre I sat down quietly, relaxed, and said to myself, "I know and I believe that today the God-Self will give me a title for my manuscript." It was not until you said during the course of the lecture, "Intellect cannot solve the problems with which people are faced today," that I realized in spite of my statement of belief to myself, I was still trying to solve it on the conscious level. The title then popped into my mind "like a piece of toast out of a toaster," as you have often said. I knew it was absolutely right and exactly fitted what I am trying to say in the manuscript. I was so elated I could have laughed out loud—it was so simple, "the impossible made possible," by reliance on the subconscious.
>
> I won't reveal the title—even to you, dear Dr. Murphy, but it is just what I needed to renew my enthusiasm to work on the manuscript, and I don't want to weaken that feeling nor the effectiveness of the "answer" by talking too much about it. Suffice it to say I am most grateful. Sincerely, Mrs. J. R. Arnold, Los Angeles, California 90039.

Mrs. Arnold realizes and knows that this Cosmic Healing Power works in your business or professional world just as well as it does in the healing of the body.

How a Schoolteacher Healed Her Ulcers and Achieved Promotion

The following is a letter from a schoolteacher:

Dear Dr. Murphy: For months I have been praying for a healing and for success in my profession. I attended your lecture one Sunday morning about a month ago at the Wilshire Ebell Theatre when you spoke on "How to Use the Healing Power." As you spoke I realized that I had been unconsciously rehearsing my aches and pains and constantly rehearsing my troubles in the class room and with my principal. I was in the habit of criticizing and blaming the pupils, the parents, and the school authorities. I was also in the very bad habit of criticizing and condemning myself for being ill. I felt sure that my desire to be promoted would fail.

It seemed to me that throughout your lecture you were talking to me, yet I knew that you didn't know I was in the audience of 1300 people. Before your lecture was over, I became aware of the fact that I had been actually squandering the treasures of life within me on negative thoughts and imagery that were positively destructive.

I followed your instructions and began to identify with the Infinite Healing Presence within me, and I affirmed frequently, with deep understanding, the meditation you suggested:

God hath not given me the spirit of fear; but of power, and of love, and of a sound mind! [2] *I have a firm, unwavering faith in God as my bountiful, ever-present Good. I am vitalized, energized, healed, and made whole. Promotion is mine now. I radiate love and goodwill to all my pupils, my associates, and to all those around me, and from the depths of my heart I wish for all of them peace, joy, and happiness. The intelligence and wisdom of God animate and sustain all those in my classes at all times, and I am illumined and inspired. When I am tempted to think negatively, I will immediately think of God's healing love.*

I want to thank you for that lecture. I have had a perfect healing of my ulcers, and I have received a promotion and

[2] II Timothy 1:7.

have established a very harmonious relationship with all my associates and pupils. I understand the reason for all this is the fact that the children in my class subconsciously picked up my new mood of love, goodwill, and confidence as I constantly exude vibrancy and joy. Yours very truly, E.R.B.

How a Five-Word Formula Healed Epilepsy

The following interesting letter was received while I was writing this chapter:

> Dear Dr. Murphy: I listen to your radio program every morning. This morning I heard you read a letter from a radio listener about a wonderful healing which she had. May I tell you about the marvelous healing which I had?
>
> It has been nearly four years since I have had an attack of epilepsy! I never knew what caused the seizures. I had been taking about three or four grains of phenobarbital every day and was sleepy all the time. One morning on radio you discussed some psychosomatic research work on epilepsy, saying that *suppressed emotions, intense hatred, and hostility toward parents often were causative factors.* This was true in my case. I began to pray systematically, pouring out love and goodwill toward my parents until I could meet them in my mind and be at peace.
>
> Every night and morning for about six months, I repeated slowly, "*God is healing me now.*" One morning I knew I was healed as a tremendous feeling of joy welled up within me. I stopped taking the drug and went to my doctor; he conducted the usual brain tests—all were negative. That was almost four years ago.
>
> The five-word formula (*God is healing me now*) you gave on the radio sank into my soul, and I am eternally grateful. It was so true that the suppressed negative electricity had to have a negative outlet, and it was expressed as epilepsy in my case. It is true that love is the great healer. Sincerely yours, J.D.M.

The Law of Belief and How to Use It

The law of life is the law of belief, and belief could be briefly summed up as a thought in your mind. To believe is to accept something as true. Whatever your conscious, reasoning mind accepts as true engenders a corresponding reaction from your subconscious mind which is one with Infinite Intelligence within you.

Your subconscious mind works through the creative law which responds to the nature of your thought, bringing about conditions, experiences, and events in the image and likeness of your habitual thought-patterns, proving the truth stated so succinctly in the Bible: *As a man thinketh in his heart, so is he.*[3]

SUMMARY

Points to Remember

1. The Infinite Healing Presence which made you can heal you and make you whole. There is only one Cosmic Healing Power and that is God within you.
2. A woman healed her crippled hand by calling on the Infinite Healing Presence, and It responded and acted according to her belief; the so-called impossible became possible.
3. A cyst on the ovary and an abscess in the kidney were dissolved by a woman who believes that God's love can dissolve everything unlike Itself. If you can think a tumor into being (and you can), it is possible to "un-think" it.
4. In spiritual healing, you must understand what you are doing and why you are doing it. True spiritual healing is the synchronous union of your conscious and subconscious mind, scientifically directed by Cosmic Power.

[3] Proverbs 23:7.

5. A tailor's eyesight returned when he entered into the spirit of forgiveness and affirmed feelingly and knowingly: *By day and by night I am seeing more and more of God's love, light, truth, and beauty in every person and in every thing.*
6. A miracle is something that happens naturally when one brings in a higher Cosmic law than those which man has heretofore known.
7. A writer healed her frustration by realizing that Infinite Intelligence knows the answer and that she was receiving it *now*. The answer popped into her mind like toast pops out of a toaster.
8. A teacher ceased rehearsing her aches, pains, and difficulties, and turned to God. She claimed promotion, peace, health, and harmony, and her desires were granted. The Bible says: *He shall call upon me, and I will answer him* . . . (Psalm 91:15). God is the Infinite Intelligence within you which responds to your thought.
9. There is a five-word magic formula for healing: *God is healing me now.* A woman applied it and was healed of epilepsy of ten years standing.

6

How to lead a successful life and gain promotion

Every man's life from beginning to end is a plan of God. The famous German philosopher, poet, and dramatist Johann Goethe said, "Life is a quarry, out of which we are to mold and chisel and complete a character." The Bible says, *I am come that they might have life, and that they might have it more abundantly* (John 10:10).

You are here to lead a full, happy, and glorious life. You are here to release your hidden talents to the world, to find your true place in life, and to express yourself at your highest level. When you find your true place in life, you will automatically become a tremendous success. You will be perfectly happy, and health, wealth, and all the blessings of life will follow.

Your success or failure in the art of living a glorious and wonderful life depends on the nature of your habitual thinking and on your true desire to transform and to recast substantially those patterns of thought. Get a new perspective, a new approach, and a new vision about life, God, yourself, and the universe.

Remember, there is a right way and a wrong way to think,

speak, act, sing, drive a car, to imagine and feel, and to conduct a business or a profession. Learn how in Cosmic Wisdom to think right, feel right, act right, be right, do right, and pray right, and all your ways will be ways of pleasantness and all your paths will be lanes to happiness and peace.

How a Woman Received Specific Guidance

The following is a letter from Mrs. Vera Radcliffe, a distinguished organist, an artist who has tremendous faith in the Cosmic Power:

> Dear Dr. Murphy: Three years ago I was travelling in Kashmir, India. I became acquainted with a prominent jeweler in Srinigar, recommended by mutual friends. Since I could buy there much below the American market, I bought $800 worth of their specialties, superior to gems from any other country, such as star rubies, topaz, and sapphires. I checked the credit of said jeweler, and, getting fine reports, I left a check for the order and as I was on my way around the world by air travel, had the merchandise sent to my home.
>
> After returning home, the package did not arrive at the appointed time. Over the next year there were no answers from Kashmir to my eight polite but firm, insistent letters. Since I had exhausted my conscious directions of action, I started praying for Divine guidance and knowing that there would be a complete filling of the order.
>
> I have a friend of whom I'm very fond but whom I seldom see or talk with, since we live about fifty miles apart and our busy schedules do not fit timewise. Gene telephoned and said we hadn't chatted leisurely for over two years and would I come to lunch, so we met at a lovely restaurant centrally located for us both. After "catching up" on our activities, she asked how I liked India. I replied that I had had a wonderful time but with one flaw—that of the above transaction. She disclosed she was a close friend of the Vice President of India. He had stayed at her home as an exchange student many years ago. This all came as a complete surprise to me.

Gene volunteered to write for help in my behalf. Within five weeks all the merchandise was delivered complete—with apologies! Please accept my sincere thanks for helping me with this problem. Kind regards, Vera Radcliffe, Studio City, California.

A Hidden Talent Revealed

A young woman came to see me some months ago and said, "I'm a misfit. No one wants me. I'm a square peg in a round hole." I explained to her that each person is unique and that no two people are alike, any more than are two crystals of snow, or two leaves of a tree. God never repeats Himself; an infinite differentiation is the law of life, and there is no such thing as an unneeded person. I quoted Emerson for her, where he said, "I am an organ of God and God hath need of me where I am, otherwise I would not be here."

She asked, "What is it God wishes me to do?"

The answer is simple, and the prayer that she applied to follow God's will also is simple, direct, and to the point.

> *God reveals to me my hidden talents and whispers into my heart the thing He wants me to do. I know that God is Infinite Intelligence and is seeking expression through me. I am a focal point of Infinite Life in the same way that an electric bulb is a focal point for the manifestation of electricity. God flows through me as harmony, health, peace, joy, growth, and expansion along all lines. I recognize the* lead *which comes into my conscious, reasoning mind, and I give thanks for the answer now.*

After a few days a deep urge came to her to take a certain business course which she is now ardently pursuing, and because of her great sincerity she will undoubtedly be an outstanding success.

Living the Triumphant Life Now

In February of 1966, I gave a series of lectures at the Unity Center in New Orleans, Louisiana. A newspaper man there told me that one of his assignments at one time had been to ask a number of people the following question: "What have you to live for?" He said that some of the answers took his breath away, and that others made his hair stand on end.

Many men said in substance, "I'm here to eat, drink, and be merry, for tomorrow we die." A large percentage of the people interviewed said that they were waiting to retire at 65, and then they had plans to travel to different parts of the world. Some said that they were waiting to die, that they were good Christians and would go to Heaven and there be forever with God.

About 10 per cent of those interviewed said that they didn't know why they were here or where they were going, and that when they died it would be the end—they would become just clods of earth, and there was no future life. About five per cent were waiting for their children to grow up and get married; then they would travel and do the things they had always wanted to do. A small percentage were waiting for the old folks to die, and then they would decide what to do.

All these people were waiting for something to happen, instead of realizing that "God is the Eternal *NOW*."

Accept Health, Wealth, and Happiness NOW

NOW is the time. Countless people are continuously looking forward to the future for better times. They are constantly saying that some day they will be happy, prosperous, and successful.

The other day in a restaurant, I heard a man say to his companion that some time he would hit the jackpot and make his mark in the world. The other replied, "I hope some day I will get a healing for my arthritis." They were postponing their good and looking to the future for its fulfillment.

The truth is that all the powers of the Cosmic Mind are within.

Peace is now; you can claim that God's river of peace flows through you. Healing is now; feel and know that the Infinite Healing Presence which made you is now transforming, healing, and restoring every atom of your being. Claim that the Creative Intelligence which made you knows how to heal you, and that Divine Order governs your mind and body.

Wealth is available now—it is a thought-image in your mind. If you claim it boldly now, a new creative idea will come to you worth perhaps a fortune. Affirm: *God's wealth is now circulating in my life. I am engraving this idea in my subconscious mind, and I know that whatever I impress on my subconscious mind will come to pass.* Actually, the response of your subconscious is compulsive, and you will be compelled to express wealth. Why wait for it?

Strength is now. Call on the Infinite Power of God within you, and this power will respond, energizing, vitalizing, and renewing your whole being.

Love is now. Know and believe that God's love envelops and saturates your mind and body, and that this Divine love will be filtered through and made manifest in all phases of your life.

Guidance is now. Infinite Intelligence within you knows the answer, and It responds to the nature of your request.

Claim your good NOW. You do not create anything; all you do is to give form and expression to that which always was, now is, and ever shall be. Moses could have used a loud speaker, a radio, or a television. The idea or principle by which these are made always existed in Infinite Mind. Plato referred to the "archetypes of Divine Mind," which simply means that there is an idea or a pattern in Divine Mind behind every created thing in the universe.

How to Plan a Glorious Future for Yourself

Did you ever stop to think that if you are planning something in the future, you are planning it now? If you are fearful of something in the future, you are fearing it now. If you are thinking of the past, you are thinking of it now. The onlv thing you have to change is your present thought. You are aware of your present thought, and all that you can realize is the outer manifestation of your habitual thinking at the present moment.

The past and the future are two thieves. If you are indulging in remorse and self-criticism over the past mistakes and hurts, the mental agony you experience is the pain of your present thought.

If you are worried about the future, you are robbing and stealing from yourself joy, health, and happiness. Count your blessings now and get rid of the two thieves.

To think of a happy and joyful episode in the past is a present joy. Remember, the results of past events—good or bad—are but the representatives of your present thinking. Direct your present thought into the right channels. Enthrone in your mind peace, harmony, joy, love, prosperity, and goodwill. Dwell consciously and frequently on these concepts and claim them—and forget all other things.

> *Finally, brethren, whatsoever things are true, whatsoever things are honest, whatsoever things are just, whatsoever things are pure, whatsoever things are lovely, whatsoever things are of good report; if there be any virtue, and if there be any praise, think on these things (Philippians 4:8).*

Take this spiritual medicine frequently and you will experience a glorious future.

How a Maid Got a Wonderful Car by Writing a Letter to Herself

A maid was putting aside $3.00 a week out of her meager wages in order to buy a car. Her sister gave her a copy of one of my books, *The Power of Your Subconscious Mind,*[1] and she read it avidly. She later told my secretary, Mrs. Jean Wright, that she sat down one night and wrote a letter to herself in order to impress her sub-

[1] *The Power of Your Subconscious Mind,* Joseph Murphy, Prentice-Hall, Inc., Englewood Cliffs, New Jersey, 1963.

conscious mind with the idea of a car. The gist of her letter was that she gave thanks to God for the lovely car that she now had and that she accepted it gladly and rejoiced that it was fully paid for and functioning perfectly. She placed the letter in a desk drawer and marked the envelope, "My answered prayer. Thank you, Father."

The sequel is interesting. On the following Sunday she went to her church, and, during a conversation with one of the ushers, she commented on his new and beautiful Cadillac. He said, "I want to sell one of my cars. Do you know anyone who wants a good car?" she replied, "I do, but I have only $45.00 saved." The usher said, "That will be all right; I have no storage space for it. Take the car for $45.00"—which she did. Her car has been functioning perfectly for over two years. Before, she had figured it would take her about three years to accumulate funds for a down payment on a car. But she had claimed the car *NOW*.

A car is a Divine idea in your mind, and if all the motors in the world were destroyed by some holocaust, an engineer could design another and we would have millions of motors in a few months. The idea and the principle of an automobile are in Infinite Mind, which is within you.

How an Eight-Year Old Boy Received a Gift He Wanted

Shakespeare said, "All things be ready if the mind be so." The Bible says, *Say not ye, There are yet four months, and then cometh harvest? behold, I say unto you, Lift up your eyes, and look on the fields, for they are white already to harvest* (John 4:35).

These two quotations apply only to your mental and spiritual world. As previously indicated, all things are present in Infinite Mind as ideas, mental patterns, and principles.

A boy about eight years of age was brought to see me by his mother. He had been disobedient, obstinate, and rebellious. It seemed that all his playmates had dogs, most of which were Irish terriers, and he was constantly crying for a dog. He deeply resented his mother and father for their unwillingness to grant him a pet dog. His mother had told him that dogs were dirty animals and she would not have one in the house, but that when he grew up and was eighteen he could have one and take care of it himself.

The little boy could not understand why he had to wait ten years for a dog. At my request, the mother left him alone with me for half an hour or so, and he poured out all his troubles. I told him to picture the dog he wanted and in his imagination to fondle it and stroke it every night prior to sleep, just to feel that there was a dog right there in the room with him, and that he had his arms around the dog. I suggested that he do this every night.

The boy's birthday occurred a few weeks later, and his grandfather visited him, presenting him with a $3,000 check to be deposited for his education later on—and in addition he brought him a puppy, an Irish terrier! The grandfather was received with *éclat* by the whole family, and all opposition to the dog melted in the burst of approval of the two gifts.

This little boy did not have to wait ten years for a dog after all. He had collapsed time by the intensity of his thought when he entered into the joy of having a dog *now*. The Book of Proverbs says, *Hope deferred maketh the heart sick . . .* (Proverbs 13:12). Your harvest is ready *now* in your mind. Ready your mind to receive your good *now* without further postponement.

How Cosmic Love Responded to a Widow

I recently interviewed a widow who said that she had been praying three years for a husband, but that she had never met the right man. In talking to her I discovered she had been erecting obstacles and barriers in her own mind, because her silent thought was, "I would like to marry when I retire; then I could travel and be free to enjoy life with my husband." She had been projecting marriage into the future and was defeating her own purpose.

I explained to her how to collapse time. I suggested that she feel the ring on her finger at night prior to sleep, sensing the tangibility and naturalness of it; moreover, it would mean to her psychologically that the marriage to the ideal man had already taken place and that she should rejoice in the reality of the marriage now.

She did this every night for a week, entering into the feeling and delight that would be hers if she were already married. At the end of that time her son's Boy Scout master, whom she had met socially, proposed to her, and I had the privilege of performing the marriage ceremony.

This proves that you can realize in your mind the desire of your heart without procrastination. This woman's mental picture and the *feel* of the ring were conveyed to her subconscious mind, and the Infinite Intelligence resident in her subconscious responded and brought both of them together. She called her subconscious her "invisible matrimonial agent"—and rightly so!

How a Promotion and an Enormous Increase in Salary Were Obtained

An engineer once told me that he was trying hard for promotion, but that since others in the organization had priority over him in years of service, he supposed he would have to wait perhaps for several years. This engineer also had a pilot's license and flew planes for research purposes from Los Angeles to New York and other cities. I asked him how long the flight to New York would take. "Oh," he said, "by jet, less than five hours," i.e., he collapsed over 3,000 miles of space into five hours or less. The old horse and wagon might have done it in eighteen months.

Our mathematicians and scientists are pointing out that time and space are one and correlated, and that when we collapse time we collapse space. I told him frankly that he promoted himself, but first he would have to remove the barricade and stumbling blocks which were there in his own mind, preclusions such as "Others are ahead of me," "I'll have to wait," etc.

For about five minutes morning and evening he quieted his mind and imagined his wife saying to him, "Darling, I'm delighted about your promotion and your increase in salary. It's wonderful!" He felt her embrace and sensed the reality and naturalness of her voice, her gestures, and her expressed joy.

In a few weeks' time, he realized his cherished goal. At this writing, he is working on his ideal project, doing top secret work, with an increase in prestige and an enormous increase in salary.

How a New Lease on Life Was Secured

Recently a man phoned me from Texas and elaborated on his many troubles, and he ended up by blaming God for all his reverses. I explained to him, however, that the universe is one of law and order, and that God among other things is Principle or Law, and if man breaks a law, he will suffer accordingly. It is not a question of punishment by an angry God. On the contrary, it is an *impersonal* matter of cause and effect. If a man misuses the law of mind, the reaction will be negative, but if he uses the law correctly, it will help him and heal him and restore his soul. I instructed him over the phone on how to become a free-flowing channel for Divine Life and gave him the following prayer to use frequently:

> *I am a clear, open channel of the Divine, and Infinite Life flows unobstructively through me as health, peace, prosperity, and right action. I am constantly releasing new, creative ideas and am setting free the imprisoned splendor within.*

This man has received a new lease on life and has told me that he now is just beginning to live. He added ruefully, "I stopped blocking my good. I have taken my foot off the hose, in a manner of speaking, and the waters of Life are flowing abundantly into my life."

He has learned to relax and let go, and he has ceased pressing the weight of his negative mentality upon the infinite pipeline of life. Consequently he immediately began to receive the blessings of life.

How to Experience Happiness and Success in Your Life Right Now

There is only one Cosmic Power animating the entire universe. God is Life and that is your life now, but this Life-Principle may be directed constructively or destructively, because you have the ability to choose and to decide.

When you tune in with the Cosmic Power and let It flow through you harmoniously, peacefully, and joyously, and when you think right, feel right, and act right, your life will be one of unalloyed happiness and success along all lines—right here and now.

You are using the Life-Principle destructively whenever you indulge in fear, regret, remorse, or in any form of negative thinking. All resentment, bitterness, hostility, spiritual pride, self-will, criticism, and condemnation of others are especially disastrous methods of misapplying the Life-Principle.

When we cohabit mentally with thoughts of fear, anger, hate, or jealousy, our life force gets snarled up in our subconscious mind, in the same way as if you put your foot on the garden hose, thereby blocking the flow of water. In like manner, the negative emotions which are dammed up in our subconscious come forth as all manner of diseases, both mental and physical.

The Joy of Overcoming All Your Problems with Cosmic Science

You are here to reproduce all the qualities, attributes, potencies, and aspects of God. Inasmuch as this is the true reason for your existence, it behooves you to have a wholesome dissatisfaction with anything less than complete harmony, health, and peace of mind.

An inquietude or restlessness regarding frustration, lack, and limitation should become a great incentive to you, enabling you to overcome all difficulties through the Cosmic Power within you.

Your joy is in overcoming. Problems, difficulties, and challenges of life enable you to sharpen your mental and spiritual tools and enable you to tap your hidden power and to release the treasures of the infinite storehouse of riches within.

Whatever you desire is already available as a thought in your mind. Claim what you want, and feel its reality. Infinite Mind is spaceless and timeless. Cease limiting yourself. Remove all stumbling blocks which are in your mind and enter *now* into the joy of the answered prayer.

. . . *Lift up your eyes, and look on the fields; for they are white already to harvest* (John 4:35). The harvest is the fruit of Cosmic goodness.

SUMMARY

Profitable Thoughts

1. You are here to lead a full, happy, and glorious life.
2. There is a right and a wrong way to think, speak, act, sing, drive a car, to imagine and feel, and to conduct a business or a profession. You must know the difference.
3. Call upon the Supreme Intelligence within you, and It will whisper into your heart the thing God wants you to know and to do.
4. Health is now. Wealth is now. Power is now. Love is now. Guidance is now. God is the Eternal *now*.
5. If you are planning something in the future, you are planning it now. If you are thinking of the past, you are thinking of it now.
6. You can write a letter to yourself, thereby impressing

your subconscious mind with your idea or desire. Mark your letter "My Answered Prayer." It works!

7. All things be ready if the mind be so.
8. Your subconscious is the "invisible matrimonial agent." If a woman will feel the imaginary ring on her finger prior to sleep, sensing the tangibility and the naturalness of it, her subconscious will respond with the tangible answer.
9. Every man promotes himself. Imagine that a loved one is congratulating you on your marvelous promotion. Keep it up! Perseverance with this picture pays fabulous dividends.
10. The universe is one of law and order, and if a man misuses the laws of mind he will suffer accordingly.
11. Negative emotions snarled up in the subconscious mind come forth as all manner of disease and malfunction.

7

The greatest secret of the ages

You will never be truly effective in prayer until you awaken to the greatest of all truths and the secret of all the Bibles in the world. It is expressed in our Bible in Deuteronomy 6:4: *Hear, O Israel: The Lord our God is one . . .* which means: *Hear* (understand), *O Israel* (illumined or awakened man): *the Lord* (the Lordly Power, the Supreme Power) *our God* (our Ruler, the Cosmic Power) *is one* (One Power—not two, not three, not ten, not 1,000—just *one*).

The Truly Effective Prayer Which Transforms Your Life

God is, and all there is, is God. This Presence and Power are within me, flowing through me as harmony, health, peace, joy, right action, abundance, true expression, and inspiration. I am a clear channel for the Divine, and I know that as I think and feel these truths I will experience all the blessings of life. I make a habit of this prayer, and wonders happen in my life. This establishes a new cycle of consciousness which transforms my life.

Keep Your Eyes on the Cosmic Beam and Move Ahead in Life

Cease going around in circles! Keep your eyes on the beam of God's glory and move forward in the Light. Cease going through life like a stereotyped machine repeating the same old clichés, thinking the same old way, and reacting in a mechanical way.

When your car becomes old, you get a new one. For much the same reason, from time to time, you get a new suit of clothes, a new home, or perhaps a new and better office. What about acquiring a new vision, a new image of yourself, right now? Expect the best, look forward with anticipation to a most glorious future, believe it is possible, live with the new image of yourself, and you will experience the joy and thrill of the fulfillment of your dream.

What are you aware of today? What is the level of your perception? You are looking at members of your family, you see your wife, your husband, your children, and you hear them speak. Yet there are many things in your home of which you are not aware. You know if you turned the dial of a television set or a radio, you would hear voices and music which are here now. But are you aware of the Divinity within you, which can heal, inspire, lift you up, reveal your hidden talents to you, and literally work miracles in your life? This untapped Power is always there. Begin *now* to use It!

How a Space Scientist Gets Answers in Space Research

Dr. Lothar Von Blenk-Schmidt, an air space scientist and member of the Rocket Society who lives near me, told me that whenever he has a problem in the engineering department relative to space projects, he stills the wheels of his mind and affirms quietly: *Infinite Intelligence within me casts light on this project,* and suddenly his mind is illuminated with the answer, which sometimes appears as a graph in his mind. Infinite Intelligence within him responds to his requests.

Get a New Image and Become What You Want to Be

You must move mentally from the old patterns of thought and in the realm of your own mind dwell on the way you want things to be. If you want to go to San Francisco, you obviously must leave Los Angeles. Likewise, if you want to be a happy man, a joyous man, a prosperous man, and a successful man, you must leave behind you as a closed book your old grudges, peeves, negative thinking, and self-denunciation and get a new self-image.

Picture yourself as the man you want to be. Be faithful to the new image, and it will sink down, by a process similar to osmosis, to your subconscious mind where it will gestate in the darkness, and after a while it will come forth in your experience as the joy of the answered prayer. You will become a new man in God and go forward from glory to glory.

How You Can Achieve Great Things

Direct your mind joyously, positively, and definitely in the direction of advancement, promotion, achievement, and accomplishment. Move mentally and spiritually to establish in your mind a new residence wherein you may live in the mental atmosphere of achieving, accomplishing, and fulfilling your heart's desire. Your assumption that the new image of yourself is real, and your sensing the reality of it in your mind, will cause your idea to jell within you and thereby become a living fact.

Don't go home or to work the same way every day; don't read the same old newspapers all the time; don't talk the same old way. Avoid the clichés of your everyday speech. Get a new set of friends, go home by a different route—you may perceive opportunities and values you never saw before! Think about everything and every person from the standpoint of the Cosmic Power, and you will achieve and accomplish mighty things.

How to Banish a "Jinx"

Last year I interviewed a man who had gone bankrupt a short time previously; he had developed ulcers and high blood pressure and was, as he said, "in a mess." He believed that there was a curse following him, that God was punishing him for his past sins, that God had it in for him, and that he was now reaping his just deserts. All these were false beliefs in his mind.

I explained to him that just as long as he believed there was a jinx following him he would suffer, for the simple reason that man's beliefs take form as experiences, conditions, and events. I gave him the following prayer which worked miracles in his life; it will do the same for you:

> *There is but one Creator, one Presence, and one Power. This Power is within me as my Mind and Spirit. This Presence moves through me as harmony, health, and peace. I think, speak, and act from the standpoint of Infinite Intelligence. I know that thoughts are things, what I feel I attract, and what I imagine I become. I constantly dwell on these truths. Divine right action governs my life. Divine Law and Order reign supreme and operate in all phases of my life. Divine guidance is mine now. Divine success is mine. Divine prosperity is mine. Divine love fills my soul. Divine wisdom governs all my transactions. Whenever fear or worry come into my mind, I affirm immediately: "God is guiding me now," or, "God knows the answer." I make a habit of this, and I know that miracles are happening in my life.*

He prayed out loud in this manner five or six times daily, and at the end of a month his health was restored and he was taken in as a partner in a growing concern. His whole life had been transformed. The new idea enthroned in his mind became his master and compelled him to express the riches of life.

Ideas are our masters. We are controlled and governed by ideas. Enthrone in your mind the Divine ideas suggested to this man, and watch wonders begin to happen in your life.

Her Belief in One Cosmic Power Healed Her

Recently I talked to a woman who was being treated with ultrasonic therapy and aspirin for arthritis. She had been praying, but she said, "Every time I begin to think of health, harmony, and peace, the thoughts of incurability, pain, and deformity fill my mind, and I can't think of health."

The reason for this was that since infancy she had been conditioned to a belief in the incurability of many diseases, and since she believed that arthritis is crippling and cannot be healed, she seemed to lack the capacity to deliberately, incisively, and decisively choose health, harmony, and wholeness. She was able to neutralize this attitude, however, after she followed my instructions.

I explained to her that the first thing she had to do was to disabuse her mind of a belief in two powers, one causing sickness and the other determining the degree of health. She actually believed in these two powers and failed to perceive the simple truth that the cause of all sickness, poverty, misery, and suffering in the world is the rather prevalent belief in them which causes the mind to be doubleminded and unstable in all its ways.

She decided to get back on the beam, so to speak, and, bringing her mental field of vision into focus, she affirmed boldly:

> *I believe once and for all that there is but one Cosmic Power, which is all wholeness, beauty, and perfection. I know and believe that the greatest secret in life is to know and to believe in the one Power which is infinitely good and perfect. I consciously claim that the healing love of this Power which created me is now dissolving all deposits in my body which do not belong there. I am*

a temple of the Living God, and I glorify God in my body.

As she continued in this prayer, her faith in the one Power increased and her belief in an evil power gradually diminished until her conviction of the one Cosmic Power reigned supreme in her mind. She continued with ultrasonic therapy, and gradually her limbs became supple. All the calcareous deposits characteristic of arthritis were eliminated, the edema subsided, and her body became a channel for the one Presence and the one Power which always moves as beauty, love, and peace. Her enthronement of the one Power in her mind and her belief that It was functioning in her body neutralized everything that contradicted this in her subconscious mind.

Let Your Head-Knowledge Become Heart-Knowledge

A woman once told me about her wonderful philosophy of life. She had written a thesis called *A Way of Life*, all of which was very sound scientifically and true spiritually. Yet her personal life was chaotic: she had been divorced four times (her age was 25), she was an alcoholic, and she was unable to pay the rent.

I explained to her that her theories, postulates, and philosophical niceties must become embodied and made manifest in her experience; otherwise they were meaningless. In other words, her "head-knowledge" had to become "heart-knowledge"; that is to say, she had to assimilate and appropriate these truths until they became a living part of her, in the same way as a piece of bread becomes a part of her blood stream.

Her thoughts, theories, and ideas—while lovely—were not demonstrated in her body, her experience, or her character. The only way she could demonstrate her theories was to put them into practice in her life, which is exactly what she had failed to do.

I gave her a pattern of prayer, and as she regularly and systematically filled her mind and heart with these truths, she became completely transformed. This is the prayer she used:

God is love, and His love fills my soul. God is peace, and His peace fills my mind and body. God is perfect health, and His health is my health. God is joy, and His joy is my joy, and I feel wonderful!

The scientist puts forth a theory or a hypothesis, but before it is generally accepted as a scientific fact it has to be validated objectively on the screen of space; otherwise it remains merely a theory. The thought must be made flesh, which means it must be embodied in your experience. You must demonstrate your religious beliefs in all areas of your life.

Why Doesn't God Do Something About War, Crime and Disease?

A man recently said to me, "If God exists, why doesn't He stop war and crime and wipe out disease?" This same cry is on the lips of many people, and it has been so for ages. Often you hear people cry, complain, whine, and say, "Why does God allow me to suffer so? Why did God visit this sickness on me? I am so good, I follow all the rules and tenets of the Church, and yet I am suffering. Why?"

The answer is rather simple, I am glad to say, and that is that God (the Cosmic Power), which is Infinite Mind, Infinite Intelligence, and Infinite Life, indwells each man, and every time man thinks, he is using the creative power—for good or for evil. Ralph Waldo Emerson said, "Man is what he thinks all day long," and the Bible says, *. . . as he thinketh in his heart, so is he . . .* (Proverbs 23:7).

Occasionally I visit people in the hospital, and some say, "Why did this happen to me? I hate no one." Others say, "God is punishing me. Why?" The explanation is the cure, in most cases: If you think good, good follows; if you think evil, evil follows.

Down through the ages theologians have brainwashed and hypnotized people into believing that evil was caused by a devil with hoofs and horns, bat-like ears, and a pointed tail which stings—a sort of hideous monster that tempts us to do evil. There is no such being; it is a figment of the imagination and a projection of the distorted, twisted imagination of man.

When we are very young, our minds are impressionable and we are not capable of indulging in abstract reasoning; therefore, we accept these weird pictures which are suggested by our parents and others, just the same as if you hypnotized a person and said to him, "You now see a snake crawling across the room." His subconscious [1] would automatically accept that suggestion, and he would see the thought-form of a snake. To him it would seem to be real. This is why you read that Martin Luther threw an inkwell at the devil, which was nothing more nor less than a projection of his own thought caused by the illusions and delusions of his mind mind regarding good and evil.

The Superstitious Origin of the Belief in Two Powers

When we were children, we thought only in pictures or mental images, and, not knowing better, we projected these images of a God and a Devil. We envisaged God up in the heavens and the Devil down below somewhere in hell, not realizing that in actual fact we create our own heaven and our own hell by the way we think, feel, and believe.

Primitive man in the jungle attributed pleasure to the gods and all pain and suffering to evil spirits or devils of his own creation. Prehistoric man realized that he was subjected to forces over which he seemed to have no control. The sun gave him heat, but it also scorched the earth. The fire burned him; the thunder terrified him; the water flooded his lands; and his cattle were drowned. His understanding of external powers consisted of primitive and fundamental beliefs in many types of Gods.

From this crude reasoning, he proceeded to supplicate the intelligences of the winds, the stars, and the waters, hoping they would hear him and answer his prayer. He proceeded to make offerings and sacrifices to the gods of the wind and the rain. Primitive man

[1] See *The Power of Your Subconscious Mind* by Joseph Murphy, Prentice-Hall, Inc., New Jersey, 1963.

divided the gods and genii into beneficial and malignant powers. Hence, the universality of these two characters in all systems of religion. Today we have a hangover from these age-old superstitious beliefs.

Your Freedom to Choose Health, Happiness, and Prosperity

When I was a very young boy, and before I had the ability to think abstractly, I was told that God was a beneficent, kindly old man up in heaven with angels all around him, that the streets of heaven were paved with gold, and that if I would be good I would some day go to heaven, but if I was bad I would go to hell.

Man is a free being, and he has the freedom to become a cutthroat, a gangster, a murderer, or a healthy, happy, joyous, and prosperous man dedicated to God, to his country, and to the world. *Man is not compelled to be good.* He has freedom to be good or to be bad. If he were compelled to be good, there would be no freedom; man would be an automaton with no free choice or will.

All of our suffering, pain, and misery are due to our ignorance, misuse, and misapplication of universal laws and principles. Our ignorance of the laws of life must yield to the wise use of both these laws and those of the universe.

There is only one Creative Power, but It is called by many names, such as God, Allah, Brahma, Reality, Cosmic Power, etc. In the absolute sense this Power moves as unity, harmony, peace, beauty, order, rhythm, and love; but man has freedom to use this Power constructively or destructively, as he chooses. *Behold I set before you this day, a blessing and a curse; A blessing if ye obey the commandments . . . and a curse if ye will not obey . . .* (Deut. 11:26-28).

There is only one Cosmic Power, and I would like to insist that you can never really experience true happiness and true peace of mind until you acquire a deep and abiding conviction in the one Presence and the one Power operating in your life.

Your real freedom from want, fear, and sickness comes with

your new awareness, conviction, and absolute belief in the one Power in which there are no divisions or quarrels. There cannot be two powers; one would cancel out the other, and there would be chaos everywhere. Become an open channel and let God's love, harmony, and peace flow through you at all times.

The Knowledge That Gives You Peace, Harmony, and Answers to Your Problems

God is Infinite, and Infinity cannot be divided or multiplied. The Bible says, *I form the light, and create darkness: I make peace, and create evil: I the Lord do all these things* (Isaiah 45:7).

This Bible quotation dramatizes and portrays lucidly that there is but one Power which you can use to give you light, which means that when called upon consciously by you, Infinite Intelligence will cast light on any problem. You create darkness when you say, "I'm stymied; I'm blocked. There is no way out. It's hopeless." By your conscious decree, when you have adopted this mental attitude, you are simply saying: "Infinite Intelligence does not know the way out," and according to your belief you live in the darkness and confusion created by your ignorance or misuse of this Cosmic Power.

You create peace by dwelling mentally on whatsoever things are true, lovely, noble, and God-like. You create evil in your experience by thinking negatively, viciously, and destructively. In other words, ask yourself this question: "How am I using this Power?" When you use It constructively, harmoniously, and according to Its nature, you may call It "God in action," and when you use It destructively, negatively, and contrary to Its nature, you may call It "the Devil" or "Satan."

Good and Evil in Your Life Are Determined by Your Thought

All the forces of nature may be used two ways. How can any man on the face of the earth become a new man and experience a new birth, or an inner transformation, except he understands the simple truth expounded here? Remember, you can never really grow and advance spiritually until you come to the absolute conviction that there is but one Power. *I am the Lord: that is my name: and my glory will I not give to another, neither my praise to graven images* (Isaiah 42:8).

You can use electricity to vacuum the floor, to illuminate your home, or to kill someone. You can use water to drown a child or to give it a bath. The same wind that drives a ship onto the rocks will also carry it to safety. You can use nitric acid constructively in various chemical and industrial processes, or you can use it to blind someone. You can use atomic energy to drive a ship across an ocean or to destroy cities, towns, and people.

Good and evil are in the mind of the individual—they are nowhere else. Whether a thing is good or bad depends on the use to which it is put, and in the final analysis, that is determined by the thought of man.

You Become What You Contemplate

Believe nothing you do not understand. Put it on the shelf of your mind and say to yourself, "My Higher Self sheds light on this subject." Every belief tends to manifest itself. For example, if you believe you have to come back again and again to this plane to expiate for your sins, you are definitely placing yourself in chains of bondage and thralldom, thereby inhibiting your spiritual growth.

Think big from now on. Enlarge your vision. Contemplate freedom, peace of mind, abundance, and goodwill to all. You become what you contemplate. Your new self-image has its own mathematics and its own power of self-expression; there is nothing to oppose you, so there is no cause for anxiety or worry.

Your Heavenly Vision

Turn your eyes to the Cosmic Power within and keep yourself on the beam of God's glory, and you will rise higher and higher every day. You will go from power to power, from octave to octave, and from glory to glory. God keep you on this path now and forevermore.

SUMMARY

Ideas Worth Recalling

1. The greatest secret of the ages is to know there is only one Power, God—not two, not three or more, just *one*.
2. The truly effective prayer is voiced while realizing that God's Presence flows through you as harmony, health, peace, joy, right action, abundance, true expression, and inspiration.
3. Keep your eyes on the beam of God's glory and move forward in the Cosmic Light.
4. If you have a problem and can't solve it, affirm quietly: "Infinite Intelligence within me casts light on this project."
5. Picture yourself as the person you want to be. Be faithful to the new image, and your subconscious will bring it to pass.
6. Think from the standpoint of Divine Principle about everything, and you will accomplish and achieve great things.
7. Ideas are our masters. We are controlled and governed by ideas of our own making.
8. Consciously claim that the healing power of God which created you is now dissolving all disease deposits in your body. Let God's love heal you.

9. Head-knowledge is not enough; it must become heart-knowledge to be truly effective.
10. Every time a man thinks, he is using the Creative Power. Use It for Cosmic good in your life.
11. Man creates his own heaven and his own hell by the way he thinks, feels, and believes.
12. You have the freedom to choose health or sickness, wealth or poverty. You are a choosing, volitional being. Choose health, harmony, and happiness as your Cosmic birthright.
13. When you use the one Cosmic Power constructively, It is called God. When you use It negatively, men call It the devil or Satan. All the forces of nature can be used two ways. Whether a thing is good or bad depends on the use to which it is put, and that is determined by the thought of man. The key to the degree of happiness in every man is his thought.
14. Our good and evil depend on the way we use our mind.
15. Every belief tends to manifest itself. Believe only that which is lovely, noble, and God-like and it will come into your experience.
16. Focus your attention on the great truths of God—this is the Heavenly Cosmic vision.

8

How to make right decisions with Cosmic Power

All successful men and women possess one outstanding characteristic, and that is their ability to make *prompt* decisions and to *persist* in carrying those decisions through to *completion.*

A distinguished industrialist once told me that in his fifty years' experience in dealing with men and women in the commercial and industrial fields, he found that all those who failed had one characteristic in common: it was that they hesitated to make decisions, and they vacillated and wavered. Further, when they did make decisions, they were not persistent in adhering to those decisions.

The Power of Decision

The power to decide and to choose is the foremost quality and the highest prerogative of man. Man's capacity to choose and to initiate what is chosen reveals his power to create as a son of God.

I have had a letter from a young man, dramatizing the power of decision. He came to a decision in his mind to acquire a Volkswagen automobile, knowing that as he decreed it with feeling, his deeper mind would respond in ways he knew not of.

How His Power of Decision Won Him a New Car

The following abbreviated letter is a marvelous example of making a clear-cut decision and persisting mentally in that decision.

> Dear Dr. Murphy: I came to a decision to purchase a car. I did not have the required amount of money. I decided to trust my deeper mind, and I dismissed the problem from my mind, knowing my subconscious had the answer.
>
> On April 8, a Friday night, a friend asked me if I would go to a teen-agers' fair, and I decided to go on Sunday night. A car was being given away that night. I had a chance of 35,000 to one to win. My name was selected, and I won the dream car of my mind!
>
> I know the reason I got the car was *my trust and faith in my deeper mind to solve the problem of a car.* As I continue to use the truths of God, my life is now in complete harmony. I would like to thank you for opening my eyes to this Cosmic Power. Seeing you each Sunday gives me what I need to go through the week. Your thoughts and words are giving me and my family a better life. Thank you sincerely, Phillip Brenot, West Los Angeles, California. P.S. You have permission to publish this.

How Courage to Decide Transformed a Life

A young woman once told me that she felt lonesome, baffled, and frustrated because she couldn't decide whether or not she should marry. Her mother was very domineering and objected to every young man in whom she was interested. This young woman had lost all initiative and power of decision with resultant loneliness and frustration.

At my suggestion, she began to make one decision after another for herself, whereas previously her mother had made all decisions for her. She decided to get an apartment for herself and to furnish it. She came to a decision to purchase her own clothes. She decided to take up dancing, swimming, and golf. She got into the habit of making all decisions for herself. She finally decided to marry a wonderful man without consulting her mother or anybody else, but just following the dictates of her own heart. She discovered that it's never too late to start making decisions and to live your own life in a wonderful way.

Remember, it is never too late to bring order to a disordered mind or to disordered affairs by coming to logical decisions and letting those decisions stand.

How Power of Decision Brought About a Miraculous Healing

The following letter shows the faith of a woman in her own mental processes and in her ability to decide, knowing that her mind is one with the Cosmic Mind of God.

> Dear Dr. Murphy: A few years ago I had a serious automobile accident. The doctor said he had never seen a neck and back broken in so many places and he doubted I would live.
>
> I decided I would live and be healed by the Power of God. I knew that all the Power of the Godhead would respond to my decision, as I had heard you say many times that it is done unto you according to your decision. I asked your Prayer Ministry for prayers, and I claimed frequently that the Infinite Healing Presence was making me whole and perfect, and a marvelous healing followed.
>
> I had been told that I would have to wear a body and neck brace for several months and perhaps a year. I wore the brace only a few weeks, and there is nothing wrong with my neck and back now. My heart is full of gratitude. I

know it is done unto you according to your decision. I decided to be healed, and the Cosmic Power responded accordingly. Signed, E.D.

How a Pharmacist Makes Right Decisions

As I was talking one day with a prominent pharmacist, he mentioned that life, with its business and professional complications and its attendant confusion, often makes decisions difficult, but that he had mastered what he believed to be the ideal method of arriving at the right decision and the correct thing to do.

He said that his favorite Biblical quotation is *Be still, and know that I am God* (Psalm 46:10). Then he added, "I dwell on the fact that God indwells me, and I focus all my attention on the Infinite Intelligence within me. I imagine that God is answering me. I relax and let go completely and feel myself completely immersed in God's quietness and stillness. Inwardly, as clear as crystal, the answer pops into my mind, and it is always right for the occasion."

This pharmacist has evolved a wonderful technique for receiving answers to problems and for coming to the right decisions with true Cosmic Power. Thomas Carlyle once said, "Silence is the element in which great things fashion themselves."

An Effective Prayer for Right Decision

This is a prayer which I have given to thousands of men and women for guidance in making decisions. They have received marvelous results and have been blessed in all their decisions.

> *Whatever I need to know comes to me from the Cosmic Power within. Infinite Intelligence is operating through me, revealing to me what I need to know. I radiate love, peace, and goodwill to all mankind in thought, word, and deed. I know that what I send out comes back to me a thousand-fold. God in me knows the answer. The perfect answer is made known to me now. Infinite Intelligence and Divine wisdom make all de-*

cisions through me, and there is only right action and right decision taking place in my life. I wrap myself in the mantle of God's love, and I know the Divine right decision is mine now. I am at peace. I walk in the Light, full of faith, confidence, and trust. I recognize the lead which comes into my conscious, reasoning mind. It is impossible for me to miss it. God speaks to me in peace. Thank you, Father, for the answer now.

Whenever you are wondering what to do or say, or what decision to make, sit quietly and affirm the above truths slowly, quietly, reverently, and with feeling. Do this about three times in a relaxed, peaceful mood, and you will receive the Divine impulse and you will experience the inner silent knowing of the soul, whereby you know that you know. Sometimes the answer comes as an inner feeling of certitude, a predominant hunch, or a spontaneous idea which wells up clearly in your mind, like toast pops out of the toaster. Intuitively, you will recognize the right answer, the right decision to make. Make right decisions by creative and intelligent prayer.

Logical Decisions Will Guide You

When you use the term "logical," you mean that your judgment is reasonable, sound, valid, based on the rational principle of the universe, or the way a thing is, or that which is consistent and deducible.

It is logical for you to think good, since only good can follow. It is illogical for you to think evil and to expect good, as seeds (thoughts) grow after their kind. This is a mental universe, and the mental law is always supreme. Logical decisions are always based on Cosmic Wisdom.

How a Woman Became a Stockbroker Against All Odds

A saleslady in one of the department stores in Los Angeles was interested in the stock market for many years and had become a very successful investor. This young woman took a required course at night which qualified her for employment in a brokerage house.

She had numerous interviews but was unable to get employment because, as she said, of her sex. She said to me, "They just don't want women." I suggested that she come to a decision and affirm boldly, "I am now employed in a brokerage firm with a marvelous income consistent with integrity and justice."

I explained to her that the minute she came to a decision in her mind and persisted in that decision, her subconscious mind would respond and open up the way by revealing the perfect plan for fulfillment of her ideal. I instructed her also to follow the *lead* which would come to her conscious mind.

The sequel is interesting. A strong urge came to her to advertise in the local newspaper, offering to work free for two months and pointing out that she had a large circle of friends as potential customers. She had immediate offers from three firms, one of which she accepted.

This shows that you must have faith in your ability to decide, and that when you come to a clear-cut decision, backed by faith in the powers of your subconscious mind, wonders will happen in your life and you will banish all frustration.

Decide to Accept Your Cosmic Divinity Now

People who fear to make decisions or who are afraid to make choices are actually refusing to recognize their own Divinity, for God indwells all men. It is your Divine and Cosmic right to choose and to make decisions.

You can decide to be healthy, happy, prosperous, and successful because you have dominion over your Cosmic world. Your subconscious mind is subject to the decrees of your conscious mind (with Cosmic understanding), and whatever you decree shall come to pass. The Bible says. . . . *Whatsoever a man soweth* (in his Cosmic subconscious mind) *that shall he also reap* (Galatians 6:7).

The Cosmic Power Is No Respecter of Persons

The law of your subconscious mind plays no favorites, no more so than any other law of nature. It is illogical to put your hand on a hot stove; if you do, you suffer the consequences. To jump off the roof of a high building is illogical, as the law of gravitation is impersonal and not vindictive in any way. It is illogical to believe that two and two make five. It is foolish to go against the laws of nature, the immutable rules of God's universe—the way things are.

What Happens from Lack of Decision to Decide

A man once said to me, "I don't know what to do or what is reasonable or logical, and I won't make a decision." I explained to him that he had made a decision. He had decided not to decide, which meant that he had decided to take what comes from the mass mind, in which we are all immersed. Also, if he decided not to decide, the random mind would decide for him, inasmuch as he refused to govern his own mind.

He began to perceive that it was foolish for him not to do his own thinking, reasoning, and deducing, thereby permitting the law of averages or the mass thinking of the race to make decisions for him.

He reversed his attitude and asserted positively:

> *I believe in my power, my ability, and the integrity of my own mental and spiritual processes, and I ask myself, "If I were God, what decision would I make?" I know my motive is right, and my desire is to do the right thing. All my decisions are based on the fact that Cosmic Wisdom is making all decisions through me, and therefore it must be right action.*

Following this prayer, this man has made all business, professional, and family decisions, and he is leading a glorious and wonderful life. He has better health, increased efficiency, more love, more understanding, and prosperity along all lines.

The Cosmic Power Backs Up All Your Decisions

The Cosmic Power backs up all *your* decisions. You are a self-conscious individual, and you have the capacity to decide. It is wrong to let others decide for you or to say, "I will let God decide for me." When you say that, you mean a God outside yourself. The only way God, or Infinite Intelligence, will work for you is through you. In order for the Universal to act on the individual plane, It must become the individual.

You are here to choose. You have volition and initiative—this is why you are an individual. Accept your Divinity and your responsibility, and make decisions for yourself; the other does not know best. When you refuse to make decisions for yourself, you are actually rejecting your Divinity and you are thinking from the standpoint of the weakness and inferiority of slaves and underlings.

How an Alcoholic Was Healed Through His Power of Decision

A confirmed alcoholic told me that a man once pointed a gun at his temple and told him he would shoot him through the head if he drank the whiskey in front of him, and he said, "I had to drink it. I couldn't stop; it was a compulsive act. I didn't care whether or not he shot me."

This is to point out to you that all the power of his subconscious mind was behind his decision. He actually experienced according to his decision. He subsequently reversed this decision, and at my suggestion he declared solemnly for about ten minutes:

> *I have come to a definite conclusion in my mind, and my decision is that I am free from this curse of alcoholism. Through the Cosmic Power which backs up this*

decision, I am completely free, and I give thanks to God now.

This man has not touched any intoxicating beverages in over five years and is completely free of the habit. He is a new man in Cosmic Consciousness . . . *Be ye transformed by the renewal of your mind, that ye may prove what is that good, and acceptable, and perfect, will of God* (Romans 12:2).

SUMMARY

Highlights to Remember

1. All successful men and women have an outstanding characteristic in common, and that is their ability to make prompt decisions and to persist in those decisions.
2. The power to decide is the foremost quality and the highest prerogative of man. It never fails in Cosmic Consciousness.
3. If you come to a clear and definite conclusion in your mind to possess a car, even though you have no money, if your decision is that your subconscious can and will supply the car and you believe implicitly in its power to do so, your subconscious will bring it to pass in ways you know not of. Its ways are past finding out by our conscious mind.
4. If you have been dominated by others and have permitted them to make decisions for you, start immediately to make decisions for yourself; then you will get into the habit of making all decisions for yourself. It is never too late to assert your own Divinity, which is

your capacity to make decisions for yourself. Do not abdicate your own power, privilege, and prerogative to others—this is a sign of weakness and inferiority, and a rejection of your own Divine powers.

5. You can make a decision that you will be healed through the Cosmic Power and if you persist in this decision all the Power of God will back you up. It is done unto you according to your decision.
6. In order to receive the right answer, sit serenely and feel yourself completely immersed in Cosmic quietness and stillness. Focus all your attention on the Cosmic Power within you. Invariably, as clear as crystal, the answer will come into your mind—and it is always right. It comes into your conscious mind like toast pops out of a toaster.
7. The prayer of right decision is based on the fact that Infinite Intelligence is responsive to your request. Know that Infinite Intelligence reveals to you what you need to know and that the right decision is yours now, and it will be right for the occasion.
8. A logical decision is one that is reasonable, sound, and valid, and, in addition, it is based on the laws of mind or the Cosmic principle of the universe. It is illogical to think evil and then to expect good fortune, health, or happiness.
9. There is always a solution for any dilemma; there is always an answer to every problem in Cosmic Wisdom. When stymied or blocked, decide right then and there that there is a solution, and your subconscious mind will be guided by Cosmic Mind to respond with the answer. It knows only the right answer.
10. Accept your Divinity of Cosmic Mind and decide now to be healthy, happy, successful, and prosperous. It is done unto you as you decide.

11. The law of your Cosmic mind plays no favorites. Use your mind wisely, judiciously, and constructively.
12. If you say, "I won't make a decision," you have decided not to decide, which is foolish and makes no sense. You are actually saying, "I'm going to let the random or mass mind decide for me." Decide from the standpoint of Cosmic Wisdom. If your motive is right, your action will be right. Decide for yourself now.
13. The Cosmic Power is behind all your decisions—good or bad. In order for God, or Cosmic Power, to act in your life, It must act through your thought and imagery. In order for the Universal Power to act in the particular, It must become the particular or individual. This is why you are an individual, which means you have volition, initiative, and freedom to be, to do, and to have all the blessings of life, if you choose wisely here and now.
14. You can be completely free of any destructive habit by coming to a definite decision in your mind and then knowing that all the Power of the Cosmic Presence will back up your decision. This is your freedom, your enfranchisement, and your passport to the joyous Heavens of your own mind.

9

Cosmic Power—your most powerful friend

You personally have the infallible power to direct the Cosmic Power which created the universe. It is called by various names, including subjective mind, universal subconscious mind, and the supreme power, as well as many others. This Cosmic Power stands always at your beck and call and obeys your decrees and beliefs. It is your obedient servant, and awaits only your use and direction.

Your subconscious mind [1] is your most powerful friend. It watches over you while you are sound asleep and controls all your vital processes. It never sleeps and is constantly reproducing your habitual thinking into form, function, experience, and event. Once you get the knack of directing it, you will find that it will heal your sick body or sick pocketbook or even your sick human relationships.

Cosmic Power Used to Heal a Dying Son

The following letter from Mrs. Hilda Hatcher shows how a woman believed with all her heart that if she would succeed in

[1] *The Power of Your Subconscious Mind,* Joseph Murphy, Prentice-Hall, Inc., Englewood Cliffs, New Jersey, 1963.

impregnating her subconscious mind with the conviction that her son was made whole, her prayer would be answered. She knew that God works by law and that that law is also the law of her own subconscious mind, and, furthermore, that it is done unto her as she believes.

> Dear Dr. Murphy: You are welcome to use any part of my letter, or all of it, with any modifications or alterations you deem appropriate.
>
> Some years ago, my youngest son, Charles, was stricken with polio and was given up by the doctors. I refused to accept the verdict, for I knew the healing power of God would heal my boy and I decreed that Charles would live. I knew that as I affirmed this with faith, sooner or later the idea would sink into my subconscious mind and that the miracle would happen.
>
> The doctor told me after 17 days that there was no hope and that even if Charles lived he would be in a wheel chair for life and his brain would be affected. I rejected the diagnosis and kept on decreeing, "*I know Charles is alive with the life of God.*"
>
> He was put into an iron lung. I continued to recite the 91st and 23rd Psalms and repeated, "*Charles is alive with the life of God,*" knowing that my subconscious mind must respond to my faith.
>
> One night while praying I fell into a semi-trance state, and I "saw" my boy running around in the back yard, full of life, zest, and energy. It was real and very vivid. The next day the doctor said to me, "Mrs. Hatcher, a miracle occurred during the night! Charles is out of his coma, his fever is gone, and he's asking for you."
>
> Charles was moved to the Children's Hospital to complete his recovery. There, he was given day-by-day therapy, and I was taught how to take care of him at home. Six weeks later, he *walked* into our home. A year later, while working in my home, I heard the voices of youngsters in the back yard, running, jumping, and laughing. Charles was there, just as I had seen him in my vision one year previously. He was vibrant and alive with the life of God. I know that it is done unto us as we believe. Blessings, (Mrs.) Hilda Hatcher, Los Angeles, Calif.

The letter has been condensed, but all the essential points are included. This letter demonstrates the faith of a mother who persevered to the end, while knowing in her heart that her faithful friend—her subconscious—which made the body of her son could also heal him with Cosmic Power. The Bible says, ... *Go thy way; and as thou hast believed* (with Cosmic Power) *so be it done unto thee* (Matthew 8:13).

How an Educator Used Cosmic Power to Vanquish All Obstacles

Your Cosmic Power stands ready and waiting for your use, in the same way that the tap in your kitchen stands ready to give you water. The following letter from a California educator indicates clearly how you can rise above any obstacle in human relations if you want to profoundly enough, if you believe you can, and if you use your Cosmic Power wisely, judiciously, and constructively.

> Dear Dr. Murphy: Sometimes it is hard to see Godliness in people with whom we come in contact in a business way. I felt just that way a number of years ago when I was a newly-elected superintendent of schools in a midwestern college town. One man, a board member, was the bane of my existence.
>
> I met with five board members every two weeks, and I met with the county superintendent, together with other school heads in the county, from time to time as he requested.
>
> The county superintendent had told me when he nominated me for the position that I should not stay there too long—at least "not long enough to get my feet muddy." I was also told that there were two types of school heads in the area: the professional men working with and under the direction of the county superintendent and the school board men whose cooperation with the county superintendent was

only incidental. I determined to be one hundred per cent professional.

Shortly after I assumed the position, and I believe largely because of my undeviating professional attitude, a feeling of deep animosity developed between me and the board member. Since I ran things according to the suggestions of the county superintendent and the mandates of the school laws of the state and not according to personal directions given me by the board member at odd times, the animosity continued to deepen. He really had it in for me.

Since this gentleman could sway the votes of two other board members, my professional reputation was at stake. I could get fired. Generally speaking, a fired educator goes perforce into some other line of work. I had met many such people who were selling textbooks or school supplies.

I felt like the individual who had hold of the tiger's tail. I couldn't continue as I was and I couldn't let go. Something had to give.

Since I had strong religious inclinations (I was a member of the church board as well as Sunday School superintendent) I did a lot of praying over this job. I frankly could see no good in me either.

Finally I decided there really must be something good in the man. I observed that he deeply loved his children, definitely a God-like attitude. Working from that angle, since I had children of my own, I visited him often at his place of business and asked his advice.

It was a slow process. However, over a period of months we developed a greater and greater mutual regard.

When the time arrived for renewing my contract, I was offered a generous raise to remain: the county superintendent wrote an outstanding recommendation. I thanked the board members for their offer, but declined in order to move to California.

We severed our business relationship with the best of good will. Sincerely yours, William H. Thrall, San Gabriel, California.

This is a marvelous working illustration, showing the power

of Cosmic Wisdom and love in dissolving animosity, bitterness, and hatred.

How His Cosmic "Partner" Saved Him $250,000

A man visited me as I was writing this chapter and said, "I read your book *The Power of Your Subconscious Mind*,[2] and I was deeply impressed. I know that my subconscious is coextensive with Infinite Wisdom and Infinite power. I realize it is that part of the Cosmic Power which knows all things.

"I have investigated a deal and everything outwardly seems right and aboveboard, but, as you suggested in your book, I went to sleep last night and asked my subconscious to reveal to me the answer such that I would recognize it in the morning. I woke up this morning and I feel that something is wrong about the whole situation. I can't point to anything definite, but it is a constant prompting from my Deeper Self, saying 'No!' "

He then asked me what I thought about it. I replied that undoubtedly it was his subconscious seeking to protect him from financial disaster, as his total investment in the proposed project would be a quarter-million dollars. I suggested that he not invest but rather that he drop the whole project—which he did.

I left a blank space in this chapter until I had heard from him about the denouement of the deal. I received a call from him recently saying that further investigation by private detectives whom he had hired revealed that his would-be business associates were professional confidence men who had arranged everything to appear right on the surface, but the whole project was a gigantic swindle. His Cosmic "partner" saved him $250,000.

You must learn to expect answers and to have faith in Divine right action in all your undertakings. Every Sunday a number of people leaving my public lectures say to me in substance, "I received this morning the answer I have wanted." Why did they? Because they asked for an answer and *believed* that their subconscious would respond in Cosmic Wisdom accordingly—and

[2] See *The Power of Your Subconscious Mind,* Joseph Murphy, Prentice-Hall, Inc., Englewood Cliffs, New Jersey, 1963.

they were in the right place at the right time to receive an answer. Invariably it is something that is said during the lecture that clears up the matter in their minds.

How Cosmic Power Saved a Life Through a Dream

You do not dream with your conscious mind. When you dream, your conscious mind is asleep and is creatively joined to your subconscious. Your subconscious mind dramatizes its contents during sleep, and it may present many symbolic pictures and incongruous situations.

Dreams are the television series of your Deeper Mind. There are all kinds of dreams, including dreams of pre-vision, where you see an event before it happens objectively and which may apply to yourself or to others. The dream may reveal the fulfillment of your desire or it may be a warning to avoid a tragedy.

The following letter will illustrate how your subconscious or Cosmic partner seeks to protect you.

> Dear Dr. Murphy: I have read your recent book *Your Infinite Power to Be Rich,*[3] and I think it is the finest thing I have read along that line. It is simply marvelous.
>
> I would like to tell you how my subconscious saved my life. A few months ago I was scheduled to go on a plane, and the night prior to my departure I read in my dream the headlines of a newspaper which told about a plane disaster in which all the passengers were killed.
>
> I woke up, startled and full of trepidation and a sense of foreboding. I found my wife staring at me, and she said, "Did you have the same dream that I had?" Lo and behold, she told me the same dream—exactly as I had seen it and in all its details! I cancelled the trip and the plane went down and all lives were lost, exactly as I saw in the dream.
>
> Truly, what you said in another of your books, *The Power of Your Subconscious Mind,*[4] is true: . . . *The Lord* (your

[3] *Your Infinite Power to Be Rich,* Joseph Murphy, Parker Publishing Company, Inc., West Nyack, N. Y., 1966.

[4] *The Power of Your Subconscious Mind,* Joseph Murphy, Prentice-Hall, Inc., Englewood Cliffs, New Jersey, 1963.

subconscious mind) *will make myself known in a vision, and will speak unto a man in a dream.* (Numbers 12:6).

Subconscious Instinct

There is an intelligence in the insect which staggers the imagination of man. Within the insect are the instinct for protection and an intelligence which cares for it, causing it to reproduce after its kind, and seeking to preserve it.

How Your Subconscious Instinct Works

Infinite Intelligence permeates all of nature. For example, there is an intelligence in the lowly atom because it takes two atoms of hydrogen combined with one atom of oxygen to create a molecule of water. There are life, intelligence, and power in all forms of nature. This subconscious intelligence operates instinctively, automatically, mechanically, and mathematically, according to the nature of the thing through which it operates.

In the long history of man down through the thousands of years before he became self-conscious, man had many experiences, all of which are recorded on the memory tablets of the subconscious mind and which move through all of us as instinct. For example, although you have never seen a poisonous cobra, should you see one in the jungle, you will immediately recoil because the instinctive memory of its death-dealing fangs and poisonous nature has been recorded in your subconscious mind.

All of man's experiences, plus his beliefs, fears, opinions, and convictions, are called "race belief," in which there is much that is extremely negative and in which there is also some good.

How the Cosmic Mind Forms Habits and Reveals Answers

You can choose now to insulate yourself from the "race beliefs," thereby freeing yourself from the experiences of the race mind Habits are formed by the reaction of your subconscious mind by repeating certain thoughts or actions over and over again until they become impressed permanently on your subconscious; then there is the automatic response of your conscious mind. This is sometimes called "second nature," which simply means the principle of action and reaction which is cosmic and universal throughout all nature. The action is your thought, and the reaction is the subconscious mind responding according to the nature of your thought. The great servant is your subconscious mind which acts as habit.

You teach your mind to drive a car, to operate a typewriter, to play the piano, or to walk or to talk. After practicing or learning how to drive, you gradually condition your subconscious mind so that when you get into your car you decide where you want to go and your obedient servant takes charge and drives you there. Our organist, Mrs. Vera Radcliffe, can play marvelous music blindfolded. It is her subconscious mind activating the playing and not her conscious mind. You, similarly, can type out letters with great speed, you can think abstract thoughts, and you can speak intelligently without purposefully directing your tongue.

This is the Infinite Intelligence in all people. It can reveal to you seemingly hidden knowledge and tell you what you need to know about life's problems. It sees all and knows all. You can ask a question of your subconscious mind, and it will answer you as an idea welling up from the depths of yourself.

How Immediate Guidance Was Received from Cosmic Mind

The following letter is from a listener to my daily radio program.

> Dear Dr. Murphy: My wife and I had to make a serious decision. We found ourselves confronted with a problem which completely baffled us. We got conflicting advice from three different attorneys, and the advice we received from our pastor was wholly unsatisfactory. We were frustrated.
>
> Then my wife said, "Let us seek guidance," and I got an intense urge to turn on the radio. I heard you speaking on Divine guidance. We listened for the fifteen minutes of the

broadcast, remaining silent. You closed your broadcast by saying, "Infinite Intelligence leads and guides you in all your ways and reveals to you the answer which comes into your conscious, reasoning mind, and it is impossible for you to miss it. Infinite Intelligence knows only the answer, and because It knows you know. Believe the answer is yours in the same way you believe the sun will rise tomorrow morning, and all power to you!"

I want you to know that an answer to our problem came out of the blue, which proved to be the perfect solution. An idea welled up spontaneously in my wife's mind to see an old friend who gave us the answer, and our problem was solved. Truly, there is a principle of Cosmic guidance which you so eloquently spoke about yesterday morning in your broadcast. Sincerely, J.H.

Expect an Answer

This man's letter demonstrates how an answer may come to you. Often your subconscious directs you to look in a book to find the answer or to turn on the radio at the right moment to get the right answer, or you may hear the answer at a lecture to which you were invited.

Your servant (subconscious mind) is always helping you, directing you, and guiding you. Its intimations, urgings, promptings, and desires are always lifeward.

SUMMARY

Points to Remember

1. You have the power to direct the Cosmic Power which created the universe. When you love you are using a

part of the Infinite love within you. Your subconscious mind obeys your decrees and beliefs.

2. Your subconscious responds to your measure of belief. A mother believed with all her heart that her boy, who was in an iron lung, would live. Her prayer was answered, and the doctors called it a miracle.
3. In strained relationships with another person, radiate harmony, peace, love, and goodwill to the other, and you will activate the majesty and wisdom of your Cosmic subconscious which will result in a harmonious relationship.
4. When you ask your subconscious for guidance, often it responds to you as an inner feeling that something definitely is wrong about the particular situation or project; it is an inner, silent knowing, saying to you, "Don't touch it."
5. Your subconscious may at times warn you in a dream of impending danger, enabling you to avoid a tragedy. This is why the Bible says, . . . *The lord* (the law of your subconscious mind) *will make myself known in a vision, and will speak unto a man in a dream* (Numbers 12:6).
6. There are life, intelligence, and power in all forms of nature. Your subconscious intelligence operates instinctively, automatically, mechanically, and mathematically in all forms of nature and in animals. Man can direct his own life, as he can choose his thoughts and direct his imagination.
7. The memory and experience of the entire race since the dawn of time is inscribed indelibly on the memory tablets of your subconscious mind. This is called "instinct," which, together with all the beliefs, fears, opinions, and convictions of humanity, is termed "race mind."
8. Your subconscious is the seat of habit. Habits are formed by repeating a thought pattern or an action over and over again until the subconscious assimilates the pattern,

and then there is an automatic response from the subconscious. This applies to good habits as well as to bad habits. Your thought is action, and your subconscious reacts according to the nature of your thought.

9. Often the answer to your question is given in various ways. Your subconscious "partner" may direct you to a particular book for the answer or cause you to turn the radio dial and hear a broadcast which answers your problem. Believe that your subconscious will answer you in the same way you believe the sun will rise tomorrow morning, and according to your belief (a thought in your mind) is it done unto you.

10

The Cosmic vision of "healthy-mindedness"

The Bible says: *Where there is no vision, the people perish...* (Proverbs 29:18). It also says: *I will lift up mine eyes unto the hills, from whence cometh my help* (Psalm 121:1).

I am sure your vision is directed toward perfect health, harmony, and peace of mind. You can be absolutely convinced that the mental picture to which you remain faithful will be developed in your subconscious mind and be made manifest consequently in your experience.

Ask yourself as you read this chapter: "Where is my vision now?" The answer lies in what you are mentally focussed on right now; in other words, that to which you are directing your attention in your thought, feeling, and imagery. You go where your vision is, for attention is the key to life.

The Psalmist says: *He maketh my feet like hinds' feet, and setteth me upon my high places* (Psalm 18:33). Herein lies a great psychological secret which teaches all of us how to climb to the high places of true vision. The hind is famous all over the world for her sure-footedness, for there is a perfect coordination of her rear feet with her front feet. This enables her to climb the highest mountain, as her rear feet track exactly where she has planted her forefeet. This teaches each of us that the mind and heart must agree on one's personal aims for perfect health, abundance, and security. When the words of our mouth and the feeling in our heart agree, nothing will be impossible unto us and we shall reach the heights.

How Cosmic Vision Paid Fabulous Dividends in Health, Wealth, and Happiness

Mr. Frederick Reinecke, who is a brilliant engineer, wrote me as follows:

> Dear Dr. Murphy: I have become a transformed man in the past ten years. Prior to that time, I used to have constant colds and the flu every winter, and I used all manner of cold remedies; furthermore, I suffered frequently from hemorrhoids, stomach ulcers, and spinal troubles—all due, I now know, to stress, strain, suppressed rage, and resentment.
>
> I began to listen to your lectures about ten years ago, and I learned the law of my mind and realized what I was doing to myself. I began to focus my attention on perfect health, on harmony in my business, and on Divine right action in all my affairs. I forgave myself for harboring negative thoughts and decided to forgive and to release mentally all my relatives and all those who irritated me. I used this prayer frequently:
>
> *I walk in the light that all good is mine. I am always at peace, poised, serene, and calm. The fullness of God is made manifest in all departments of my life. I constantly vision for myself and for others, health, prosperity, and all the blessings of life.*
>
> I have been in perfect health for ten years, and wealth has flowed to me, filling up all the empty vessels in my life. I am deeply grateful. Frederick Reinecke, President; FEBCO, Incorporated, Sun Valley, California.

I know Fred Reinecke very well, and I am aware of the many difficulties and challenges he has met and overcome in his life through his vision of the Cosmic Power and his attention to the sunshine of a better life. To use his own words: "The old Reinecke died. The new Reinecke, with his alignment with God, was born." Mr. Reinecke's high vision has paid him fabulous dividends in health, wealth, and happiness.

How the Cosmic Power Revealed a Life-Saving Idea

Dr. Lothar von Blenk-Schmidt writes as follows:

> Dear Dr. Murphy: After reading your latest book, *Your Infinite Power to Be Rich*[1] (which in my opinion is another outstanding work of yours), I would like to tell you what happened to me some time ago. As you know, we have an extremely high number of freeway accidents caused by people driving in the wrong-way exit. After focussing my attention frequently on the matter and visioning a solution whereby people's lives could be saved, one morning as I awakened the idea of a device and the complete outline of it popped spontaneously into my conscious mind. This device will stop cars from entering the freeway ramp the wrong way and will save countless lives.
>
> As required, I filed my invention with the company where I work; after some study, I was informed that the idea was not feasible and was also impossible. I received a legal release. All my associates—engineers and physicists—laughed at the idea and considered it doomed to failure from the beginning, just a waste of time, effort, and money.
>
> I applied for a patent on April 24, 1966, and a patent was granted from United States Patent Office on August 9, 1966, under Number 3,266,013, Title: 'Freeway Safety Device.' All my claims were accepted, and no changes were made or infringements cited. Many firms are now jumping on the bandwagon, saying, 'We always knew you had something wonderful!'
>
> I put a foundation under my castle in the air. I had a vision of saving the lives of countless numbers of people from becoming physical and mental wrecks, and I looked forward

[1] *Your Infinite Power to Be Rich*, Joseph Murphy, Parker Publishing Company, Inc., West Nyack, N. Y., 1966.

with faith and confidence to the realization of my vision; my subconscious mind responded and gave me the answer. You have my full permission to publish this letter in your forthcoming book. Sincerely, Lothar von Blenk-Schmidt.

Dr. Schmidt, a research engineer and physicist presently engaged in space research, is a profound student of the laws of mind. In his letter he points out vividly and lucidly that he saw only one thing: the open road to the realization of his goal and victory. His vision was truly on "healthy-mindedness" for all.

How the Cosmic Power Healed a Cripple

In November 1966, I gave a series of lectures at the New Orleans Unity Society in New Orleans, Louisiana, on my latest book, *Your Infinite Power to Be Rich* (referred to earlier in this chapter). One of the men present told me that some months previously he had been crippled with arthritis and was unable to bend his knees. One evening a holdup man came in and pointed a gun at his head and said, "Kneel down behind the counter!" He replied, "I can't. I'm crippled with arthritis." The bandit said, "I'll give you ten seconds to kneel down or I'll kill you!" The man told me, "I bent my knees easily! Gradually, I have gotten much better. My doctor says that the calcareous deposits are all gone, and the suppleness and mobility of my joints have returned. How do you account for it?" This is a good question.

I explained to him that there was a principle of healing involved there and that if he had understood the principle and applied it, he could have been healed prior to the hold up. When he bent his crippled knees at the point of the pistol, it stands to reason that the power to bend his knees, to walk and to run, was always present within him—even though he believed otherwise. The Cosmic Power is always available, but man's fears, false beliefs, and misbegotten concepts hold him in bondage.

In this instance, he was healed by the Cosmic Power within him, as quite obviously the pistol had no power to heal. He had been crippled for several years, he said, and was unable to bend his knees, and all his attention and vision were bound up in his pain, aches, limitation, and physical handicap. Suddenly, his attention was taken off himself and his ailment, and his vision was directed to save his life at all costs, as self-preservation is the first law of life. Immediately, he released the miraculous Healing Power that neutralized and dissolved everything unlike Itself.

Cease describing your symptoms and being completely engrossed and absorbed in your ailment or you will make matters worse. Deflect your aim in life and raise your sights. Let your vision be on perfect health and vitality, and you will begin to release the Cosmic Power instantaneously. Train and discipline yourself to raise your sights and to maintain the high vision. Chant the beauty of the good, and keep looking up.

The Miracle of Faith and Love

In Minneapolis I lectured every evening for a week on *The Miracle of Mind Dynamics*[2] at the Church of Divine Science conducted by Dr. Vernon A. Shields, Minister-Director. I talked to one of his parishioners who told me that about six months previously, he had been stricken with coronary thrombosis and had collapsed on the street. An ambulance took him to the hospital, where he received immediate treatment. The doctor frankly told him, however, that it looked hopeless and that probably he would not live more than a few hours. On hearing this, the man said to the doctor, "I'm going to live; my two sons need me. I won't die! I must live! I have so much to do! I love my children, and they need me!"

He continued, saying, "I felt as though a spiritual transfusion of some kind flowed through me. At the end of ten days in the hospital, the cardiograph showed a normal heart. I got well because I rejected the negative prognosis and identified myself with the Healing Power of God."

[2] See *The Miracle of Mind Dynamics*, Joseph Murphy, Prentice-Hall, Inc., Englewood Cliffs, New Jersey, 1963.

This man had a constructive vision, and he held on to it. His inner vision and love for his children released the Cosmic Power of God and transformed his whole being, making him strong and perfect again. Faith in God and love for his children welled up within him, and the miracle of healing followed.

How a Mother's Vision of Healthy-Mindedness Worked Miracles for Her Son

Mrs. Betty Reinecke writes as follows:

Dear Dr. Murphy: My son, Fred, Jr., was constantly being criticized by our relatives, and he felt rejected, unwanted, and insecure. Frequently he would go into a depressed state and become rebellious. I began to picture him as he ought to be: joyous, happy, and free. Every night and morning, I made it a habit to relax my mind and body, to close my eyes, and to mentally envision Fred saying to me, 'Mom, I am happy and at peace. I feel wonderful, and I am receiving wonderful marks in school.'

As you have suggested so often, I made a joyous mental movie of my own in my mind, knowing that sooner or later Fred would confirm objectively what I pictured and felt subjectively. Shortly after this mental process, my son, Fred, decided to come to your lectures, and he began to imagine better grades in school, forgave himself for harboring negative, resentful thoughts toward his relatives, and regularly poured out love and goodwill to all of them.

A miraculous transformation has taken place in his life since he started redirecting his thoughts and imagery. He is happy and free, and he is thrilled over the way he has overcome all the obstacles and challenges he met. He has frequently confirmed objectively what I imagined him telling me subjectively.

He is a communications specialist in Thailand now and is serving his country nobly and proudly. He has benefited greatly from the many years of listening to you. Today he is full of faith and confidence in the God-Power within which responds to his deep conviction. It is wonderful! Gratefully, Betty Reinecke (Mrs. Frederick Reinecke).

You Can Rise in New Confidence Through the Cosmic Power Within You

In the third chapter of Acts is a wonderful story of a lame man at the gate of the Temple Beautiful, and Peter and John said to him ... *Look on us. Peter ... took him by the right hand ... and leaping up, he stood, and he began to walk, and entered with them into the temple, walking, and leaping, and praising God.* (See verses 4 through 8.)

Peter means faith in God, and *John* means love of God or the good. Have faith and confidence in the qualities and attributes of God within you, and love in the sense that you wish all the blessings of life for all men everywhere.

Lift up your gaze and get a new vision of yourself, and the Power of the Almighty will respond and enable you to rise, walk, and leap, and praise the miracle-working Cosmic Power within you.

SUMMARY

Important Thoughts in This Chapter

1. Let your vision be on perfect health, harmony, and peace of mind. Your vision is what you are giving attention to right now—what you are thinking about and picturing in your mind.
2. Ask yourself what you are mentally focussed on. Give your wholehearted attention and devotion to the noble, the wonderful, and the God-like concepts of life.

3. Your mind and forgiving heart must be in perfect alignment in order to get an answer to your prayer. Forgive yourself for harboring negative thoughts, and release mentally all those for whom you have any grudges or animosity. Wish for them all the blessings of life.
4. Picture your loved one telling you what you long to hear. Feel the joy of it all; make it vivid and real, and you will experience the joy of hearing the loved one telling you objectively what you hear and picture him telling you subjectively.
5. If you visualize and meditate on something wonderful for other people, your subconscious Cosmic Power will supply you the appropriate idea according to the nature of your request. It comes to you in unexpected ways.
6. You can build castles in the air, but be sure you put a foundation under them. Know that it is possible to bring about the execution of the Cosmic idea enthroned in your mind. Look forward with faith and confidence to the realization of your desire.
7. The Infinite Healing Presence is within you, and as you claim, feel, and know that the Healing Presence that made you and created all your organs can heal you, the miracle-working Power will respond according to your faith. This Healing Cosmic Power is always available, waiting for you to call, and It will answer.
8. Let your vision and attention be concentrated on perfect health and vitality, and you will begin to release the Healing Cosmic Power instantaneously.
9. Reject all negative predictions and cease carping against the bad, but rather claim the beauty of the good and identify yourself with the Cosmic Healing Power.
10. You can project a mental movie in your mind night and morning, hearing a loved one tell you what you long to hear. As you remain faithful to your mental picture,

it will be developed in the darkness of your subconscious and will come to pass.

11. Lift up your gaze and get a new vision of yourself as you want to be, and, like the lame man in the third chapter of Acts, you will be lifted up and you will walk, leap, run, and praise God, and you will experience the joy of the answered prayer.

11

The magic of faith in the Cosmic Power

Faith is a way of thinking. It is a constructive mental attitude, or a feeling of confidence that what you are praying for will come to pass. Faith as mentioned in the Bible does not refer to faith in any particular creed or religious persuasion; rather, it refers to faith in the laws of your mind which each of you can learn and apply.

Actually, you do everything by faith. You learn how to drive your car by repeating certain thought processes and muscular actions over and over again; after a while the driving becomes habitual, and there is an automatic reflex action from your subconscious whereby you find yourself driving your car without any conscious effort. In the same way you learned to walk, to dance, to type, to swim, and to perform many other activities.

For example, the farmer has faith that the seed he has deposited in the ground will grow after its kind. He has faith in the laws of agriculture. The electrician has faith that electricity will respond to the laws of conductivity and insulation, and he knows that it flows from a higher to a lower potential. Edison had an idea for a phonograph; he proceeded to bring it to pass by having faith in the execution of his invisible idea.

You have faith when you know that thoughts are things; what you feel, you attract; and what you imagine, you become.

Everyone Has Faith in Something

It is true that everyone has faith in something. The atheist has basic faith in the laws of nature, in the principles of electricity, chemistry, and physics. What is your faith? Let it be faith in all things good, a joyous expectancy of the best, and a firm belief inscribed in your heart that the Cosmic Power will lead you out of your difficulty and show you the way. Have a firm conviction in the healing power of God to make you physically and mentally whole. This faith will enable you to walk over the waters of fear, doubt, worry, and imaginary dangers of all kinds.

How Faith Paid Off a $25,000 Mortgage

As I was writing this chapter, I had an interesting conversation with a man from North Carolina whose life, by faith and courage, became amazingly different. He said, "After reading your book, *Your Infinite Power to Be Rich*,[1] I decided to take God as my partner, realizing that God is my Higher Self, or the Living Spirit within me. I talked to this Invisible Presence and affirmed, *You are my Partner, and I am going to ask you to guide, direct, and prompt me in all ways. We are now a team, and we can't fail.*"

This young man told me that he had lacked the courage to go into business for himself, but after praying in the above manner and affirming his union with the Divine Presence within him, a new wave of strength and confidence had surged up within him. He made a small down payment on a restaurant, and with the magic qualities of faith and courage he paid off the mortgage of $25,000 by the end of the first year. His relatives had told him that he didn't have a chance and that he would go broke, but he realized deep within himself that nothing could defeat him, as

[1] *Your Infinite Power to Be Rich*, Joseph Murphy, Parker Publishing Company, Inc., West Nyack, New York, 1966.

he possessed the priceless quality of faith in the Cosmic Power within him to watch over him in all his ways.

His secret was that he held on tenaciously to the idea that his Senior Partner would advise him and prosper him until his subconscious mind had absorbed the idea and had brought it to pass.

How to Increase Your Faith in Your Power to Realize Your Desire

You increase your faith when you realize that your desire, idea, plan, or dream is real, even though invisible. To know with certainty that the idea is real, that it is a fact of your mind, gives you faith and enables you to rise above confusion, negation, strife, and fear to a place of conviction deep in your own heart.

How a Playwright Used the Magic of His Faith in the Cosmic Power

During a recent lecture tour in Minneapolis, a young man visited me at the hotel and showed me a play which he had written. I read part of it and found it to be extremely interesting and fascinating, but his conflict was that everywhere he submitted it for approval and acceptance he received a rejection slip. He added that he was getting a rejection complex.

I suggested to him that he change his attitude. He had to realize that the idea of the play came out of his mind; that the idea was as real as his hand; and that if the manuscript were lost or destroyed, he could write another one—that ideas were like seeds and that we don't give vitality to the seed: it has its own method of unfoldment within it and all that we do is deposit seed in the ground, knowing that it will grow. We may water and fertilize the seed, but we can't make it grow.

He began to look upon his play as a Divine seed (idea) just the same as the idea of a radio is real in the mind of the inventor or the idea of a new skyscraper is real in the mind of an architect. It is not an idle daydream. An idea has form, shape, and substance in another dimension of mind. He began to affirm boldly:

> *Infinite Intelligence gave me the idea of a play; it is a good play and will inspire and uplift mankind. I accept the fact now that the Creative Intelligence within me which originated this idea opens up the door for its perfect fulfillment. I know that the law of attraction is working for me, and I am now attracting the right people who will accept, promote, and produce this play. I release my request to my Deeper Mind, and I know that just as a seed deposited in the soil grows, expands, and unfolds, so will my desire deposited in my subconscious mind bring about the perfect unfoldment and manifestation of my play.*

In due course I received a wonderful letter from this young man, saying that he met a motion picture director from Beverly Hills on the local golf course and that during the game he discussed briefly the play he had written. The director was interested and asked to read it. He was fascinated by it and has entered into an agreement with the young man to produce the play and has made arrangements for the talent, the producer, and all other essentials. This is the great law of attraction, based on faith in the Creative Intelligence and courage to act upon this faith which brings answers to your prayers.

Faith is trust. You trusted your mother when you were in her arms; you looked into her eyes and you saw love there. Your faith in the Cosmic Power, which is All-Wise, All-Powerful, and All-Love, should be even greater than was your faith in your mother.

Self-Doubt and Fear Conquered Through Faith in the Cosmic Power

Some months ago I talked to a man from Wyoming who had been offered a big promotion which necessitated his moving to San Francisco. He said that he was bogged down with fear, anxiety, and a sense of inferiority, and that he felt he couldn't accept, that he was sure he would fail to live up to their expectations.

In other words, he was halted by self-doubt. He said that something kept welling up in his mind, whispering to him, "You can't do it."

I explained to him the source of that whispering, pointing out that it came from those same fears, anxieties, inferiorities, and self-doubts which had been deposited in the deeper layers of his subconscious mind, probably dating back to childhood, and that these negative thoughts dominated, controlled, and obsessed him.

I gave him a simple formula which acts as a wonderful mental catharsis and is highly effective when practiced sincerely and systematically. I explained to him that if he had a pail of dirty water, he could have clean water in a little while by continuing to pour in clean water; likewise, he was now to fill his mind with the following healthy ideas which would crowd out of his mind all the negative thoughts which were lodged there. This prayer was recommended:

> *I have absolute faith in God and all things good. I am one with the Cosmic Power, and one with God is a majority. I know that God and the universe are for me, and nothing is against me. I am absolutely fearless, as I know the thing I might fear does not exist—it has no power; it is only a shadow in my mind, and a shadow has no power. I am full of faith and confidence. I have the courage to meet all problems head on, and I overcome them through the wisdom and power of God within me. God's power is with my thoughts of good. I am immersed in the Divine Presence. God's love fills my soul and His river of peace flows through me. There is no fear in love, for love and recognition of the One Power casts out all fear. Every moment of my life I am growing in faith, courage, and confidence, and I feel the power of the Almighty flowing through me now. I am at peace.*

The mind—like nature—abhors a vacuum, and as this young man began to fill his mind with the above spiritual thoughts for about ten minutes each morning, afternoon, and night, he succeeded in washing out of his mind all the fearful, anxious thoughts which had been holding him for years in the thralldom and bondage of inferiority and doubt.

Following this technique of meditation, he accepted the promotion in San Francisco and was transformed by the renewal of his mind. His courage to meet the situation and to tackle the fears and cast them out was the first step in his triumphant journey to success, promotion, and greater income.

It was Shakespeare who said, "Our doubts are traitors,/ And make us lose the good we oft might win/ By fearing to attempt."[2]

How Faith in Cosmic Power Wrought a Health Miracle

Here is a letter from Chicago that is most striking:

> Dear Dr. Murphy: My boy was stricken with a very severe case of polio. He was unconscious in an iron lung for some time. One doctor encouraged me to pray. I read *The Miracle of Mind Dynamics*[3] over and over again. I applied the various prayers you have given in that book, and I fixed my mind on several quotations from the Bible: . . . *Before they call, I will answer; and while they are yet speaking, I will hear* (Isaiah 65:24). *Thou wilt keep him in perfect peace, whose mind is stayed on thee: because he trusteth in thee* (Isaiah 26:3). . . . *Thy faith hath made thee whole* (Matthew 9:22). . . . *If thou canst believe, all things are possible to him that believeth* (Mark 9:23). *A merry heart maketh a cheerful countenance* . . . (Proverbs 15:13). . . . *I am the Lord that healeth thee* (Exodus 15:26). . . . *What things soever ye desire, when ye pray, believe that ye receive them, and ye shall have them* (Mark 11:24). *I will restore health unto thee, and I will heal thee of thy wounds, saith the Lord* . . . (Jeremiah 30:17).
>
> I anchored my mind on these passages and on many more

[2] *Measure for Measure*, I, v.

[3] *The Miracle of Mind Dynamics*, Joseph Murphy, Prentice-Hall, Inc., Englewood Cliffs, New Jersey, 1964.

> which I got from your writings. I continued to practice saturating my mind with these passages, and on the third day I felt a deep, abiding sense of peace and tranquillity. I looked at my boy's unconscious face and I saw him smile at me, and I knew at that moment that God had answered my call.
>
> That happened four months ago. My boy's strength is gradually returning. The doctors have given me great encouragement, and I know that God is making him whole.
>
> You may use this letter, using my initials only. Sincerely yours, L.J., Chicago, Illinois.

Look at the magic and miracle-working power of the faith and trust of this woman who persevered in affirming the great truths of God when all seemed hopeless! Within you also are all the courage, faith, and confidence you need to overcome any obstacle. Draw on the Cosmic Power for all your needs, and you will discover that you have the courage and the power to conquer that so-called insuperable and insurmountable barrier.

Within You Is the Cosmic Power to Work Wonders in Your Life

You will be given courage, faith, and hope when you realize that you are a medium for the transmission of the Cosmic Power of God. Paul said: . . . *There is no power but of God: the powers that be are ordained of God.* (Romans 13:1)

For example, atomic power is of God; man may use it any way he wishes. Electrical energy is of God. The mighty gusts of wind represent the power of God. The power that turns the earth on its axis and moves the planets and the galaxies in space—all these mirror the Cosmic Power.

You are one with this Infinite Power, and you can contact It through your thought. God indwells you, walks and talks in you, and is the very life of you. Acknowledge this Power, unite with this Presence and become this moment a potential transmitter of wisdom, power, love, light, and truth.

Don't ever say, "I can't overcome these difficulties. I can't rise above this problem or challenge." Actually, what you are saying is that God can't meet the problem or reveal the solution, which is atheism and a denial of Omnipotence.

Affirm boldly: *I can do all things through the Cosmic Power which responds to me and strengthens me in all ways.* Wonders will happen in your life.

How the Cosmic Power Healed a Minister's Acute Migraine Attacks

A minister friend of mine in San Francisco visited me some months ago and discussed his awful migraine headaches, adding, "Often, I get one of these blinding, splitting headaches while giving a sermon on Sunday mornings." He stated that the special drug the doctor prescribed sometimes relieves, but sometimes brings no relief at all. "Often," this minister said, "I could scream with pain."

He asked me to place my hands on his forehead (which I rarely do) and pray with him, which I did. He said to me, "Looking at your hands, I know they are transmission belts for the healing power of God." Then he quoted from Mark 5:23 . . . *I pray thee that thou come and lay thy hands on her, that she may be made whole; and she shall live.* And from Mark 6:5 . . . *He laid his hands upon a few sick folk, and healed them.* I placed my hands on his forehead and prayed as follows:

> *Whatever is bothering you is leaving you now, and you are filled with the free-flowing, healing, harmonizing, vitalizing life of God. The Cosmic Power flows through my hands now, permeating every atom of your being, transforming every organ of your body into God's perfect pattern. His river of peace flows through you, and His ocean of love saturates your whole being. You are cleansed and made whole. God's love is touching you now, and we give thanks for your freedom.*

Both of us remained quiet for about fifteen minutes, with my hands resting on his head. We concentrated on a transfusion of God's grace and healing power. This minister felt a tremendous sensation of heat and a tingling vibration all over his body; he perspired copiously. He exclaimed, "I know I'm healed and free!" He has had no recurrence and is completely free. This is the magic of faith in the Cosmic Healing Power. . . . *Thy faith hath made thee whole* (Matthew 9:22). It transcends the power of drugs when properly applied.

SUMMARY

Some Profitable Pointers

1. Faith is not a religious persuasion, but rather a way of thinking, a constructive mental attitude, or a feeling of confidence that Infinite Intelligence will respond to your call upon It and that what you are praying for will come to pass.
2. The farmer has faith: he knows when he puts a seed into the ground, it will grow after its kind. You have faith when you know that thoughts are things; what you feel, you attract; and what you imagine, you become.
3. Everyone has faith in something. Even the atheist has faith in the laws of nature and in the principles of electricity, physics, and chemistry. Your faith should transcend these, and your faith should be in the goodness of God, in the guidance and love of God, and in the laws of your mind—all of which never change; they are the same yesterday, today, and forever.

4. Let God (your Higher Cosmic Self) be your silent Partner, and your oneness with this Force will enable you to move upward and onward for joyous living. Have the faith to take the step, and then have faith in the Cosmic Power to back you up—and It will.
5. You increase your faith when you know that the idea you are now entertaining in your mind is as real as your hand or your head. It has form, shape, and substance in your mind, and as you nourish it with faith and expectancy it will be objectified on the screen of space.
6. Ideas are like seeds deposited in the ground: they grow after their kind. You don't give vitality to the seed; it has inherent within it its own manner of unfoldment. Likewise, as you water your desire with expectancy and fertilize it with faith, you will accelerate its manifestation as form, function, experience, or event.
7. As you fill your mind with the eternal verities and life-giving patterns of Cosmic thought, you cleanse your subconscious mind of all negation, casting out all fear and making way for faith, courage, and love.
8. As you anchor your mind on certain deep, spiritual, healing passages of the Bible, such as . . . *I am the Lord that healeth thee* (Exodus 15:26); . . . *Thy faith hath made thee whole* (Matthew 9:22); and many others, you will gradually saturate your subconscious mind with Cosmic Wisdom and the miracle of healing will follow.
9. All power is of God, and God is Omnipotent. You are one with God, and you should never say, "I can't do this," or, "I'm incurable," or, "I can't solve the problem." Actually what you are saying is, "God and Cosmic Wisdom can't solve the problem," or, "God can't heal me." At that moment you are an atheist because you are denying the Infinite Presence and Power.
10. Healing by the laying on of hands and calling on the

Cosmic Power within you to heal has been practiced down through the ages. Many people have great belief and faith in the power of touch and the laying on of hands, and according to their faith is it done unto them.

12

How to get what you go after

Desire is the gift of God, or the Cosmic Power, seeking expression through you. When you are hungry, the Life-Principle supplies you with the desire for food in order to preserve you. When thirsty, you desire water; when cold, you build a fire; when sick, you desire health.

The Limitless Cosmic One within you does not desire to express Itself in any form of limitation. The desires, urges, intimations, and promptings of your Deeper Mind are always lifeward, urging you to rise, to transcend, to grow, and to manifest your desires for health, happiness, prosperity, true expression, and the fulfillment of your ideals, dreams, and aspirations. Desire is the goad which pushes you onward, upward, and Godward.

Men who had strong desires and powerful ambitions to get what they went after and to achieve their goals in life, made America what it is today—the greatest industrial nation in the world. For instance, Henry Ford desired to build an automobile. Then he had another desire still greater than the first, which was to put the whole world on wheels. The attainment of this desire gave employment to millions of people all over the world and blessed humanity in countless ways.

Desire is the beginning, and its realization is the end and the solution to your problem. Failure to realize your desire is the cause of frustration, unhappiness, and illness. To continue desiring to reach your goal morning, noon, and night over a prolonged period of time, yet failing to get what you go after, brings chaos and confusion into your life and is the cause of endless suffering—mental, emotional, and physical.

You Create Your Own Destiny

The Bible says: . . . *Choose you this day whom ye will serve* (Joshua 24:15). You can choose right now to mold, fashion, and create a wonderful future. Your thought and feeling control your destiny.

Emerson said, "Man is what he thinks all day long," and the Bible says, *As a man thinketh in his heart, so is he* . . . (Proverbs 23:7). The word *heart* in the Bible is an old Chaldean word meaning the subconscious mind—all of which means that whatever thoughts, feelings, beliefs, and impressions you have made, or have permitted others to impregnate your mind with, have a life of their own in your Deeper Mind, and they dictate, control, and manipulate all your conscious actions.

In other words and in utter simplicity, what is impressed in your subconscious is also expressed outwardly in form, function, experience, and event. Over a hundred years ago, Dr. Phineas Parkhurst Quimby said, "Man is belief expressed."

Your future, therefore, is your present habitual thinking projected on the multi-dimensional screen of space. In other words, it is your present thought and belief made manifest. Your future is your present thought patterns grown up, and this is accomplished in the same way as a field reproduces a harvest in consonance with the seeds deposited therein. All seeds (thoughts) grow after their kind.

How a Tremendous Increase in Business Was Achieved

A businessman who visited me and asked me how he should pray about his business said, "Everything is going wrong. I am going to be a failure. Everything is against me. Business is bad, and the worst is yet to come!"

I explained to him that he could reverse the entire situation by changing his thought patterns and keeping them changed. After reflecting on my advice he began to realize that he had been actually creating the conditions he feared, because his subconscious mind was constantly reproducing his habitual thinking.

I further pointed out to this businessman that he could attain success, prosperity, and inner peace—or any other good thing in life he felt entitled to—provided only that he would pay the mental price by incorporating into his subconscious mind the equivalent of it. Accordingly, every morning when he opened his eyes he affirmed boldly:

> *Today is God's day. I choose happiness, success, prosperity, and peace of mind. I am Divinely guided all day long, and whatever I do will prosper. Whenever my attention wanders away from my thoughts of success, peace, prosperity, or my good, I will immediately bring back my thoughts to the contemplation of God and His love, knowing that He careth for me. I am a spiritual magnet, attracting to myself customers who want what I have to offer. I give better service every day. I am a wonderful success in all my undertakings. I bless and prosper all those who come into my store. All these thoughts are now sinking into my subconscious mind, and they come forth as abundance, security, and peace of mind. It is wonderful!*

He started using the above prayer each morning and night,

and before the end of the month I received the following letter from him:

> Dear Dr. Murphy: This is a thank-you letter, to let you know how much I got from the interview I had with you some weeks ago. I know now that I have a right to anything that would be good for me or that would make me healthier, happier, more successful, and more useful.
>
> I had never thought that the desire for success, prosperity, and advancement was the urge of the Divine Presence in me telling me that the time has come for me to take a step up the ladder of life. Now I know that the will of God, or Life, for me is to move forward, to conquer, to rise, and to achieve great things along all lines.
>
> I am grateful for your emphasis on the truth that one of the things holding me back was that I thought I had the right to something that belonged to someone else. I now know it is dead wrong to infringe on the rights of others or to covet their goods or property or wealth. I fully realize that God can give me all the wealth, happiness, success, and opportunities in life without infringing upon the rights of anyone else.
>
> I have paid the price by constantly dwelling on the prayer you gave me, and I have noticed a big increase in business, and I feel happier and am much more cheerful. Over my desk I have placed a beautiful printed card: *In all thy ways acknowledge Him, and he shall direct thy paths* (Proverbs 3:6). The price I paid was mental acceptance and application of these truths and remaining faithful to them.

He Stopped Blaming Fate and Achieved Promotion and Financial Increase

Some months ago while staying at the Kona Inn, a favorite resort on the big island of Hawaii, I chatted with a man by the swimming pool in that lovely spot. This is the gist of what he said to me: "I came here for a week to get away from it all. My life is a daily grind. I get nowhere, even though I work hard. I am bored and jaded, and life for me is monotonous. I eat, sleep, work, and look at TV. I hate my work. There is a cruel fate holding me back, for I try so hard and yet can't make ends meet."

I gathered in conversing with this man that he actually had borrowed the money from his sister in San Francisco in order to have a week's vacation. I said to him that I would give him a simple, down-to-earth, practical formula that would completely turn the wheel of fortune and enable him to acquire wealth, success, happiness, and the full realization of his ambitions in life.

He listened avidly to the meaning of . . . *Whatsoever a man soweth, that shall he also reap* (Galatians 6:7). This means that if you plant thoughts of lack, limitation, strife, bitterness, sickness, poverty, and contention, you shall reap accordingly. You must remember that your subconscious mind is like the soil: it will grow whatever type of seed (thoughts and mental imagery) that you plant in the garden of your mind. You sow thoughts when you believe them wholeheartedly to be true; and it is what you believe deep down in your heart that you demonstrate for yourself and the world.

This man had studied Emerson's philosophy in college, yet had never understood it; he merely read it as literature. Emerson wrote in his *Essay on Fate: He (man) thinks his fate alien because the copula (connection) is hidden. But the soul (subconscious mind) contains the event that shall befall it; for the event is only the actualization of its thought, and what we pray to ourselves for is always granted. The event is the print of your form. It fits you like your skin.*

This is the formula I gave this man to use morning, noon, and night:

> *God is my Partner, and within me is the Cosmic Power to overcome any condition. I was born to win, to succeed, and to conquer. I find a great thrill in mastering difficult assignments. My joy is in overcoming and getting what I go after without infringing in any way whatsoever on the rights of others; neither do I want others to do that which they would not want me to do.*

I know that the Cosmic Presence within me can and does give me all the happiness and opportunity I need without interfering with another person's rights. I am now in touch with the Eternal Source of all supply and of all good, and the Divine Presence is bringing me peace, happiness, joy, success, prosperity, and full expression through Divine and legitimate channels. God shows me the path; I trust the Divine Intelligence within me, and I am Divinely directed and guided to my true expression in life and to the fulfillment of my heart's desires.

I wrote this prayer for him by the pool and explained that as he affirmed these truths consciously and with a deep, abiding faith, they would sink down by osmosis to his subconscious mind, and the latter—being one with the Cosmic Power and boundless wisdom—would compel him to take all steps necessary for the fulfillment of his dreams and that he would get, legitimately and through a Divine way, whatever he was going after in life.

This young man discovered that he was not a victim of cruel fate; neither was he condemned to live a life of mediocrity, poverty, and miserable conditions. He broke loose mentally from false ideas and beliefs which had held him in thralldom and bondage. The following letter from San Francisco attests to his complete mental transformation:

> Dear Dr. Murphy: I was glad to talk to you in Kona! I had never heard much about the subconscious mind and its powers. I have followed the formula you wrote out for me. I apply it three times daily, and when negative thoughts come to me, I immediately reverse them. I have discovered that what you told me is true—that the negative thoughts have gradually lost their momentum, and I am now a constructive thinker.
>
> I am now attending a course in business management and also am brushing up on my Spanish three evenings a week at the University. I have been promoted to assistant manager and have received $50 weekly increase in salary. I am

on the way up! I also am studying *The Power of Your Subconscious Mind*[1] and find it a treasure house of knowledge. Many thanks! T.I.

How a Salesman Broke Blocks to Closing Sales

A few months ago I received a letter from a man in New York City in which he stated that whenever he was about to bring a deal to the point of completion, the door seemed to jam tightly and everything would collapse at the eleventh hour. This has happened time after time when he is supposed to have the prospective client sign on the dotted line. Suddenly the client becomes ill, or his wife dies, or he meets with an accident. There is a definite appointment at an office, and the meeting falls through for various reasons, such as fog delaying the plane or sudden illness or failure of different people to show up. He asked what he could do to remove the "block."

The only "block" this man suffered from was his false belief which dominated his mind. He feared a recurrence of these episodes in his life, and what he feared most he experienced. In other words, it was done unto him as he believed in very tangible form.

In my letter to him, I pointed out that the way to harmony, success, and prosperity was within himself. He did not have to beg, supplicate, or beseech God—all he had to do was to change the stream of his mental thought and imagery. I emphasized that the affirmations I was about to write for him possessed their value only because of his thought and understanding of the laws of his mind behind them—that he could, for example, recite the Declaration of Independence and the Bill of Rights and still have little or no understanding of the true meaning of Americanism. Following my directions, he reversed his mind as follows:

I know that there is only One Mind which created all things, and my mind is one with that Cosmic Mind. I meditate on my inseparable oneness with this Cosmic

[1] *The Power of Your Subconscious Mind*, Joseph Murphy, Prentice-Hall, Inc., Englewood Cliffs, New Jersey, 1963.

Mind Power, and, without strain or stress or any great mental effort, I let my thoughts and mental images sink into my deeper mind; I am quiet in the assurance that these thoughts will return, fabricated and manifested by the Cosmic Mind who created the universe.

I know God is the Cosmic Mind Power, which is the reality of me. My work is God's work, for God works through me and my work cannot be hindered or delayed. The Cosmic Mind Power that created all things finished whatever It created. My work goes through to completion in Divine order. All that I set out to do reaches fruition. I entertain only those thoughts that make for my highest happiness, peace of mind, and achievement.

The above prayer, affirmed with sincerity and deep understanding of the way his mind worked, brought about a complete change. He reversed the so-called block and is now closing sales and moving forward with faith and confidence to greater and greater usefulness and accomplishments in life.

What You Can Conceive, You Can Achieve

An engineer in London, England, once said to me, "Some years ago when I first started, I failed miserably in three assignments given to me, and my supervisor said to me, 'You are fearing failure and you expect failure, and it is what you really expect deep down in your heart that you experience.' This was the turning point in my life, and I completely changed my mental attitude.

"I admitted to myself that I had faith in failure, not in success and getting things done. From that moment on, I had faith in success. My motto became *Anything I can conceive and believe possible, I can achieve.* My new attitude is the reason behind my being the director of this engineering firm."

I would suggest that you inscribe in your heart this engineer's quotation: *Anything I can conceive and believe possible, I can achieve.*

This engineer began to realize there was an Almighty Power within him, the Cosmic Giver of all gifts, which he could tap; he began to find the answers, the power, and the wisdom to accomplish things which he previously had believed to be hopeless. He looked forward to achievement, victory, and triumph, and his faith became contagious: all the men working under him likewise became imbued with the idea of success and triumph, and they also accomplished great things.

This engineer's favorite Bible quotation is: *Every valley shall be exalted, and every mountain and hill shall be made low . . . Make straight in the desert a highway for our God* (Isaiah 40:3, 4).

SUMMARY

Important Pointers

1. Desire is the Cosmic gift of God. The desires, urges, intimations, and promptings of the Life-Principle within you are Life's way of telling you to rise, to transcend, and to grow.
2. You mold, fashion, and create your own destiny. Your future is determined by your habitual thinking and self-imagery, and your subconscious mind faithfully reproduces what you think all day long. Emerson said, "Man is what he thinks all day long." Change your present thought for the better and you change your destiny for the better.
3. You have to pay the price in order to get what you want in life. The price is to learn the Cosmic laws of your mind and then to give attention, devotion, and nourishment to your ideals. The law of mind is that whatever you

give attention to and take a deep interest in, your subconscious mind will magnify and multiply and express on the screen of space. Your price is recognition of the laws of mind, complete acceptance of these laws, application of them, and deep abiding trust that they never fail. In short, your price is recognition, acceptance, and conviction of the Cosmic Power within YOU.

4. You have a perfect right to anything that would make you happier, healthier, or more successful in life, and one of the things that could hold you back on your upward journey is to think you have a right to something that belongs to someone else. It is dead wrong in Cosmic justice to infringe on the rights of others. Wish for all men that which you wish for yourself.
5. Realize the Cosmic Giver, your God-Self, can give you all the wealth, happiness, success, and opportunities in life without hurting a hair on the head of any living being in the world.
6. There is no cruel fate holding you in mediocrity, lack, or sickness. As you sow in your mind, so shall you reap. Plant orchids (beautiful thoughts) in the garden of your mind. All the experiences of your life are simply externalizing your thoughts, conscious and unconscious.
7. There is no block holding you back other than a false belief or an abnormal fear enthroned in your mind. A block is a false belief which dominates a man's mind, bringing all sorts of suffering in its train. Remove the block with a deep understanding of the laws of your mind.
8. Whatever you can conceive and believe possible, you can achieve with Cosmic Mind.

13

How to be joyous conquering obstacles to your Cosmic good

The principle of growth is cosmic and universal. You may see this marvelous principle of growth manifesting itself daily in flowers, in all vegetation, in animals, and in human beings. The seed of a tree has been known to overcome all obstacles—even to break through a rock—in order to become a tree. In like manner, you are here to overcome, to rise over all obstacles, to transcend, to grow and to rejoice in it.

You are here to master all problems, to solve all difficulties, and to experience joy, happiness, and victorious living. You are here to become mentally and spiritually aware of your transcendental powers which enable you to overcome any challenge and experience the joy, satisfaction, and thrill of victory.

The minute you acknowledge the Cosmic Power within you and accept your responsibilities for your environmental conditions, you will begin to gain control over conditions, and your thinking will begin to make itself felt.

It would be a very dull and insipid life for you if the crossword puzzle were all made out for you and all you had to do was to insert X, Y, and Z. No, the joy is in solving the crossword puzzle—this is the way you discover the Divinity within which shapes your destiny.

How Cosmic Mind Helped Sell Real Estate

Some months ago I interviewed a man who had had great difficulty in selling his property and home, as he said, because of tight money and high interest rates on banks loans. During the interview, I said to him, "There is an age-old truth which says, *What you are seeking is seeking you.* The mere fact that you want to sell your property and home indicates that somewhere there is someone who wants to buy it."

I then said to him, "Let us both pray about the matter and agree mentally that it is sold to the right buyer who wants the property, who appreciates it, and who prospers in it."

Accordingly, both of us stilled our minds, relaxed mentally and physically, and focussed our attention on the sale of the home. I said audibly:

> *Let us now agree that the Cosmic Mind knows where the right buyer is and that the Infinite Intelligence within that Mind knows where the buyer is, and It is now this moment in action, bringing both of you together. Both of us decree that the price is right, the time is right, and the buyer is right. We agree that the property and the home are now sold in the Cosmic Mind of God, as buying and selling represent an exchange of ideas in the mind of man. We decree that it is so, and we accept it completely in our minds right now.*

This was our prayer, to which he agreed completely. The sequel to this prayer is very interesting. When he went to sleep that night, in a dream he saw a man who gave him a check for the full value of the property. In his dream he said to the man, "Are you paying in full for the whole thing?" and the man said, "Yes." Then he awakened and intuitively knew that his property was sold.

Ten days elapsed, and on the tenth day the man whom he had seen in the dream came to him and bought the property, which consisted of ten acres and a home. The man paid him exactly the amount he asked for—which also coincided with this dream.

The Cosmic Mind knew where the buyer was and revealed the answer to him in a dream, which was subsequently confirmed. The ways of your Deeper Mind are past finding out. The Bible says: *. . . I the Lord* (your subconscious mind, or the Law) *will make myself known in a vision, and will speak unto man in a dream* (Numbers 12:6).

How a Businesswoman Triumphed over Obstacles

A businesswoman with whom I talked complained that her health, business, and home conditions were very bad and practically beyond her control. I explained to her that on the contrary, she must definitely and positively assume responsibility for her well-being, happiness, and all conditions of her life.

In her conversation, she said, "The first two years in business, I was a great success. Everything went fine, and I prospered. Now, everything is upside-down."

I asked her a simple question: "Do you take the responsibility for the two years of success?"

She said, "Of course I do. I worked hard and diligently, and I gave my best to the customers."

I said, "Your answer is correct, but you just can't take credit and approbation for your success in business—which you said was due to your efforts and mental acumen—and then refuse to assume responsibility for your failure, ill health, and discordant home conditions. This is a contradiction which is illogical, unreasonable, and unscientific."

She immediately perceived the wisdom of this explanation. Suddenly she began to realize that she had the privilege to use the

Cosmic Power to live a successful, happy, and prosperous life and it was her responsibility to use the Power rightly, not only for success and prosperity but also for her well-being and joyous living.

This woman had been deceived and swindled in a real estate deal and consequently was extremely resentful and bitter. This mood of lack and hostility lurked in her subconscious mind, and it became a festering, psychic wound which brought on her depression, nervousness, and financial distress.

She accepted the challenge of life and forgave the man who had swindled her, who, incidentally, had absconded with her money and gone off to South America. Her simple prayer was: *I bless him; I forgive him; I forgive myself, and I mean to forget it.* And it worked! She had discovered the Cosmic Power, which enabled her to overcome her resentment and all her ills. The Cosmic Power responded according to her conviction and belief. The daily prayer I gave her was as follows:

> *As I think wisely, judiciously, constructively, and harmoniously, I shall not reap a crop of confusion, lack, sickness, or misery. I steadfastly think of increase in my business, of joyous good health, and of vitality, abundance, and expansion along all lines. I know that these seeds fall on good ground (my receptive mind), and they yield an abundant harvest. I constantly sow seeds of forgiveness, release, and goodwill to all, and I know the Cosmic Power responds by delivering to me better conditions, larger experiences, and better situations. I maintain harmony and peace of mind.*

She nourished these thoughts daily, and at the end of three weeks there were a renewal and an expansion along all lines. Her business began to prosper, new friends came into her life, she had buoyant health, and she became cheerful and optimistic. This woman had discovered the joy of living, of overcoming, and of the buoyancy and resiliency caused by the flow of the Cosmic Power through her.

The Answer That Saved the Suicide

Last year the manager of a downtown hotel phoned me, and in a rather excited tone of voice he said, "There is a guest in the hotel who wants to commit suicide! He won't get out of bed; he won't eat, and he mumbles to the bellboys, 'I'm going to kill myself. Call Dr. Murphy.'"

I visited the man and said to him, "You can't solve a problem by killing yourself or by jumping out the window. You can solve your problem right where you are, because you personally are greater than any problem."

I did not plead with him. Instead I taught him something about himself, because from experience I have learned that often the explanation is the cure.

I pointed out to him that the act of trying to escape a problem is like running away from Los Angeles to San Francisco: you take your mind with you, and the only place your problem exists is in your mind.

I also dwelt on the fact that man is not just his body but that man can actually move outside his body; he can look upon his body from outside and also can be seen at a distance by others. I mentioned that these truths about man have been known for thousands of years: that there is no death—only life; that his body is an instrument to express mind and spirit; and that he will have bodies continuously to infinity.

I described the experiments and research work of the famous Dr. Hornell Hart, former associate of Dr. J. B. Rhine at Duke University, who has investigated numerous cases of men and their experiences outside their bodies. I told him frankly that he would meet the same problems outside the body he now had, as he would immediately put on another body much more rarified and attenuated and called the fourth-dimensional body. He would remain confused, frustrated, and perplexed in his new body, which would conform to his confused thoughts and imagery.

This man became intensely interested and absorbed in my description of experiments in India and in the United States on man living outside his body. He asked several questions and finally began to see that what he really wanted was a solution. His suicidal complex was caused by an intense desire for freedom and peace of mind. His urge to commit suicide was a sincere desire for a greater expression of life. He soon realized, though, that he could not really destroy life, mind, or spirit.

I explained to him that man is mind, and ever more he takes the tool of thought, and choosing what he wills brings forth a thousand joys or a thousand ills.

This man's death desire was a desire for a greater measure of life, for there is no actual extinction of life. The suicidal urge in this man, as I further explained to him, was a desire to destroy and abolish the obstacle or impediment, or, in the case of acute mental pain and anguish, the body.

The so-called insoluble problem which he had was not uncommon. His wife had deserted him and had run off with another man; where she went, he didn't know. They had had a joint bank account and she took all the money—stocks and bonds included. His store was burned to the ground and his insurance had lapsed. Then he turned to drink, sank into the throes of despair and despondency, and wanted to end it all.

He finally decided, however, to stand up to his problem like a man, realizing that he was born to win, to conquer, and to triumph over all obstacles. I took him out to dinner and talked to him about men who had lost everything and yet who had triumphed over disaster. He took on a new lease on life, and his constant prayer was: *I can do all things through the Cosmic Power which strengtheneth me. God's love fills my mind and heart, and God reveals to me the answer and the life abundant.*

The next day he went to work in a chemical plant and vowed to himself: *I will rise and overcome through the Power of God,* and he did. He obtained a divorce, and he eventually married an immensely wealthy woman; they are deeply in love with each other. All his losses were wiped out overnight. As a wedding gift, she bought him a new store with all the necessary equipment. He said to me, "Your explanation of man living outside his body has saved my life."

This man's inner realization of a Cosmic Power which is the very life of him enabled him to see that he is immortal. This is why the Bible says: . . . *This is life eternal, to know thee the one true God* (John 17:3).

How He Found a Way Out of the Wilderness in Vietnam

Recently I lectured at a banquet, and a young officer who had recently returned from Vietnam sat next to me. He told me that he had had to bail out of his plane as it caught fire, and he found himself in the jungle. Darkness was just setting in, and he was hopelessly lost. He said that he was seized with panic for a few minutes. Then he said to himself, "The night is here, and the dawn follows the night. I am going to find a resting place and sleep and know that the Lord is my Shepherd." He remarked, "When I did that, all fear went away and I knew I would be safe."

This Air Force officer had assured himself that the darkness breaks into flame and light. Fear was the darkness, and faith in God who would take care of him was the flame of light. The next morning he was rescued by friendly Vietnamese farmers who had come his way, hunting for game.

This man had overcome his fear, and the Cosmic Power had responded to his prayers of faith. . . . *According to your faith is it done unto you* (Matthew 9:29).

How an Executive Overcame Pressure and Excess Tension

Some time ago in San Francisco, where I gave a series of lectures on *Your Infinite Power to Be Rich,*[1] I interviewed a businessman at the St. Francis Hotel, where I was staying. During our conversation he said to me, "I'm very worried and terribly tense. The pressure is terrific; it's a rat race today. I'm losing my grip and giving way under it."

[1] *Your Infinite Power to Be Rich,* Joseph Murphy, Parker Publishing Company, Inc., West Nyack, N. Y., 1966.

I asked him this question: "Do you think that fear, pressure, anxiety, and worry are independent of your mind, or are they entities that walk about, ready to attack you and give you nervous prostration?"

He replied, "No; I know that disease, fear, and worry are not independent of my mind."

"I don't suppose you ever read the Bible, do you?" I asked.

"Indeed, I do," he said boldly. "I read passages every night before I retire."

"Well," I said, "I am going to give you one of the most wonderful and beautiful healing passages in the entire Bible, from the Book of Job: *Acquaint now thyself with him, and be at peace: thereby good shall come unto thee.*"[2]

I continued, "I know that you read the Bible, but obviously your mind is engrossed in the pressures and strains of your business difficulties. Now you must deflect your mind from the problems and vexations of the day and acquaint yourself with the Cosmic Power within you, which is universal and which is the Power that created all things in the universe. As you recognize and accept this Supreme Power and Wisdom and begin to call on It, It will answer you and give you victory and triumph over all your anxieties and perplexities.

"Don't give power to the vexations, troubles, and trials of the day; never give them power over you. Rise above them! You can go to the Cosmic Power within you and gather more strength and be replenished with wisdom, which will enable you to withstand the tribulations and contentions of the day and experience some good in all problems. Furthermore, you will grow tall in confidence and poise.

"From now on, keep your mind on God's wisdom and power; you will have peace of mind and will be more effective in all phases of activity. Practice the repetition, a thousand times a day, if necessary, of this passage: *Acquaint now thyself with him, and be at peace: thereby good shall come unto thee.* Get your mind away from your troubles and focus your attention on the Cosmic Healing Power within you!"

[2] Job 22:21.

He thanked me, and said, "I'll try your spiritual medicine."

A month passed by and I received a telephone call from him in which he said that he had used the passage from Job over and over again. He said that every time he was about to fly off the handle, instead of fretting, fussing, fuming, and fulminating, he affirmed the passage from Job. Now he has peace of mind and control over his emotions, and he is no longer victimized by strain and pressure.

SUMMARY

Steps in Truth to Remember

1. There is great joy and satisfaction in overcoming obstacles and difficulties. Furthermore, this is the way in which we sharpen our mental and spiritual tools, enabling us to discover the Divinity within.
2. When you wish to sell something, you must not dwell on all the obstacles; rather, you must agree in your own mind that it is sold to the right buyer in Divine Cosmic order. What you are seeking is seeking you, and the law of attraction will cause the right buyer to purchase it.
3. If you take the responsibility for success in your business or profession due to zeal, enthusiasm, and diligent application on your part, you cannot at the same time refuse to take responsibility for your failure, sickness, or discordant conditions. You have the freedom to use the Cosmic Power for success or failure, health or sickness, peace or pain.
4. A man who has a suicidal urge or complex is looking for an answer, freedom from his distressing problems or

acute mental anguish. He cannot escape his problem because it is in his mind, and he is mind and not just a body. If he destroys his body, he merely dons another and still remains confused and perplexed. Inform him that he lives outside his physical body, and when he leaves this body or destroys it, he puts on a rarified and attenuated fourth-dimensional body. The Cosmic Intelligence can solve his problem and set him free. When he calls on It, he gets the answer. Man is immortal; he lives forever, and there is no death. He meets his problems here, and he solves them through the power of God.

5. The dawn follows the darkness of the night. The light of God will illumine the darkness of the mind, which is due to fear, panic, or a sense of hopelessness. When lost in the mental jungle, sit still, relax the mind and body, and know that there is a Cosmic Intelligence which will watch over you, protect you, and lead you to safety. The day will break for you and all the shadows will flee away.
6. You overcome pressure and strain by staying your mind on the Cosmic Goodness within you.

14

The Cosmic Power and your future

You mold, fashion, and create your own future by your present thoughts made manifest, because, as Emerson said, "You are what you think all day long." Change your present thoughts and you change your whole life. You can direct the Cosmic Power within you and thereby control your life experiences; you can actually bring to pass the cherished desires of your heart.

Your mind is constantly in motion as thoughts, images, ideas, dreams, and aspirations come and go. Your world is constantly changing in harmony with your habitual thinking. You create for yourself success or failure, affluence or poverty, health or sickness, and peace or pain by your conscious and subconscious thought. *For as a man thinketh in his heart* (subconscious mind), *so is he* . . . (Proverbs 23:7). Any thought of yours—good or bad—if consciously accepted, is implanted in your subconscious mind, and it comes forth after its kind. This is why you create your own future.

How a New Future Was Gained Through Cosmic Healing Power

A few weeks ago I visited in the hospital a man who had undergone a major operation. He told me that his kidneys had completely given up functioning, and he asked me to pray for him, saying, "I have no future. I'm only forty, but I suppose I'm done for. What will happen to my family? Prayer is all that's left!"

I said to him that his first step was his willingness to believe that the Cosmic Healing Power which made his body and all its organs could restore and heal him. We prayed together as follows:

> *We join together now in knowing that the Cosmic Healing Power which made your body and all its organs knows all the processes and functions of your body, and the miraculous Healing Power is permeating every atom of your being, making you whole and perfect. All your organs are God's ideas, and through the power of the Almighty they are functioning perfectly now.*

After about fifteen minutes our prayer was answered and his kidneys began to function, which pleased his surgeon immensely. He is now back with his family and in perfect health, and, while conferring with me recently, he said, "My future is assured. I know that my future is my present thought grown up."

This man knows that his faith in the goodness of God, in the guidance of God, and in the creative power of his own thought will always be made manifest in his life. With this attitude, he is building a glorious future for himself full of harmony, health, peace, and abundance.

How to Assume Responsibility for the Future

Some months ago I had an interview with a man who was blaming circumstances, God, fate, bad luck, and his in-laws for his unhappy and inharmonious condition. I explained to him that all of God's riches were available to him: health, wealth, love, peace, harmony, and the Cosmic Creative Power, which he could instantly contact through the medium of his thought. I emphasized the fact that it was up to him to make his own future whatever he wanted it to be, but that he had to open his mind to receive the bounty of God. Here is the formula I prescribed for him, which I called mental and spiritual medicine.

> *First step.* Take all your attention away from old hurts, grievances, resentments, and rankling memories, and cease all self-condemnation or criticism of others.
>
> *Second step.* Take time out morning and evening to implant in your subconscious mind by regular, systematic, and constructive thinking, the mental patterns of harmony, abundance, security, success, and full expression.
>
> *Third step.* Instantly chop off the head of every thought of fear, worry, sickness, defeat, discouragement, or lack of any kind. This means that you cremate or burn up every negative thought by immediately substituting constructive thoughts, such as harmony, peace, love, joy, right action, and Divine guidance.

He undertook this mental discipline, and in a month's time he noticed a most remarkable change in his affairs—in all phases of his life. He held on to this new mental and spiritual regimen, perseveringly and unflinchingly, and discovered that as he changed his mind he changed everything. His business prospered, he sleeps better at night, he has ceased blaming others, and he is going ahead by leaps and bounds.

He said to me, "I realize that my future is my invisible thinking made visible." This is the truth that set him free.

It's Never Too Late to Change for the Better

I recently explained to a young woman that the future was in her own hands and that she could do literally anything with her life, because it was her life. The truth is that she could actually change her future.

She had thought heretofore that her future was in the hands of God and that she could not determine and create her own destiny.

She pointed out to me that her mother had always told her that God knew best and that she should take what comes and be content. I explained to her that this isn't the truth, *that she had to take the initiative*, and that God, the Cosmic Power, would not work for her except through her—i.e., through her thought and imagination.

She adopted an intelligent plan and decided to run her life as methodically and efficiently as a successful person operates his business. The following mental technique was outlined for her with the instruction that she was to affirm these truths three times a day feelingly and knowingly, and with a deep understanding that every thought tends to execute itself unless neutralized by a counterthought:

> *I am now designing and mapping out my own future. I am a child of God, and I have the capacity to think, feel, imagine, act, and react. From this moment forward I choose harmony, health, peace, abundance, Divine right action, happiness, and love. I know that these thoughts are like seeds placed in the ground, which contain the complete program for their full development. Likewise, my subconscious mind knows how to bring forth all the mental seeds I am depositing in my mind. I know my thought and feeling create my destiny, and I rejoice that it is so.*

In less than three weeks, by following the above technique and prayer process, she had added color, vitality, joy, happiness, and luxuries to her life. She subsequently married a young physicist, has gone on a trip around the world, and has experienced a wonderful mental transformation, whereby her life now has color, beauty, and love. She has discovered that the normal desires and demands of the full life can be met in an intelligent manner since she has permitted herself to become a creative outlet for the Cosmic Power within her.

She told me, "It is wonderful to know that I am a child of God and that God is releasing and expressing Himself through me and that I create my own future."

Why You Are Always Planning for the Future

You are "always" planning for the future, because if you are thinking or planning something for the future, you are thinking about it *now*. Likewise, if you are fearing delays, obstacles, and impediments to your future plans, you are thinking of these things *now*. In the Cosmic Mind there is no time or place—only the Eternal *now*. It is omnipresent, timeless, and spaceless.

There are many people who get the weekend off and look forward to blue Monday, or "black Monday," as some call it. Naturally it becomes a day of confusion and depression to them, as they had consciously decreed their future and their subconscious had responded accordingly. In all probability, they didn't even know that they had planned ahead and had thus created their future.

When you really *want* to change your future, you will come to a decision to change your thoughts, feelings, actions, and reactions. You can have a new birthday in God when you realize that your mind is a part of the Cosmic Mind and that the Cosmic Power is with your thoughts of good. You can be transformed and you can build a glorious and wonderful future by the renewal of your mind.

How a Young Woman Discovered She Has What It Takes to Face Life

A few years ago while lecturing at the Psychic Science College in Belfast, Ireland, I interviewed a young woman who said to me, "I haven't got the power to handle my troubles and difficulties or to meet and solve the problems of life. I'm divorced. I hate myself, and I'm no good."

I explained to her that her condition was simply due to habitual negative thinking, and to constant self-criticism and self-condemnation which were poisoning the springs of hope, faith, confidence, and enthusiasm and were rendering her a physical and mental wreck. In other words, she was taking mental, self-generated poisons, and she was polluting the sanctuary of the Living God within her.

I explained the laws of mind to her and pointed out that she is constantly reaping what she sows in her subconscious mind; that she was here, like everybody else, to overcome problems, challenges, and difficulties and to lead a victorious life; that she had to stand up as a child of God and meet life boldly; and to realize and utilize the tremendous Cosmic Power within her.

I told her the story of how Eddie Rickenbacker, the famous aviator, and his companions were shipwrecked and were adrift on a raft in the Pacific Ocean. He prayed for food, and a gull came and perched on his head and remained long enough to be seized and utilized for food. He prayed to be rescued, and of course he was picked up. He believed in the wisdom and power of God to take care of him and the answer came, for the Cosmic Power answers you when you call, believing.

This episode made a profound impression on the young Irish lady. I gave her the following prayer, pointing out that it was a process of reconditioning her mind and that when negative thoughts of any kind came to her mind—which they would because of her destructive habit of self-condemnation and self-demotion—she was to immediately supplant them with a spiritual thought. This is the prayer which she was to repeat out loud for about ten minutes, morning and night:

> *I am a child of God. I am a channel of God, and God has need of me where I am, otherwise I would not be here. I know I am here to express more and more of God's love, life, truth, and beauty. I am here to do my share and to contribute to humanity. I have much to give; I can give love, laughter, joy, confidence, and goodwill to all people, to all animals, and to all things in God's universe.*

I am here to stir up the gift of God within me. I realize that I am a gardener, mentally speaking, and as I sow, so shall I reap in life. Life is a mirror for the king as well as for the beggar, and whatever I give to life, life returns to me magnified, multiplied, and running over. I deposit in the garden of my mind wonderful seeds of peace, love, goodwill, success, harmony, and joy.

I forgive myself for harboring negative, destructive thoughts, and I pour out love and goodwill to all my relatives and to all people everywhere. I know when I have forgiven others, because when I meet them in my mind there is no sting; I am free.

I am constantly drawing out the fruit of the delightful seeds I am depositing in my subconscious mind. I know that my thoughts, like seeds, will be made manifest as form, function, experience, and conditions. I think on these things, and the Cosmic Power is with my thoughts of good. I am at peace

She repeated the above prayer for the recommended ten minutes in the morning and evening, realizing that her eyes saw these truths and her ears heard them, bringing the two faculties of sight and hearing into function, thereby reinforcing the power of her affirmations. After a few weeks, I received the following letter:

Dear Dr. Murphy: All of us enjoyed your lectures in Belfast; you opened the eyes of many. I would like to tell you of the change that has come over me.

I prayed the way you told me, and after a few days all the bitterness in my soul disappeared as if by magic. I joined a dancing class, and I have been promoted in the store to the position of head of my department. The assistant manager proposed marriage to me, and we are to be married in six months! I have forgiven myself and my relatives.

Every day is a new day. I know that I predict my own

future by the way I think, feel, and imagine. I am grateful. God bless you!

How a Salesman Changed His Mind and Changed His Destiny

Some months ago a young salesman came to see me. He said that he could not get along with his boss; he had failed to make his quota of sales and had lost his job; and he was out of money. He had walked ten miles to see his sister in Hollywood in the hope of borrowing a few hundred dollars. His sister refused, saying that until he stopped drinking and philandering she would not give him a penny, and she coldly turned him away.

He told me that he had slept in a flop-house and that the kindly desk man had told him to see me and that I would help him. I spent over an hour with him, listening to his sob story of grief, disappointments, and anger and resentment toward his former boss and members of his family who had refused to help him.

He said that he was a member of a church; that he religiously followed all the rules, tenets, rituals, and ceremonies; that he attended every Sunday morning—but still everything went against him. This young man also was angry even at God for letting him down.

In many instances, the explanation is the cure. Patiently and kindly, I explained to him that he could keep and follow all the prescribed rules, rituals, doctrines, and dogmas of a church from an external standpoint, and still think negatively and destructively. I told him that if he believed in a punitive God, or that God had not played fair with him, such beliefs automatically would bring trouble and difficulties on him. In other words, he becomes his own tormentor and brings failure, lack, and misfortune on himself, for it is done unto him as he believes.

I explained to him that the first step is to admit that he had been wrong; then he could be completely changed in all phases of his life. My mental and spiritual prescription for him was as follows:

"Form a mental image of yourself as a great salesman; imagine that your relatives and I are congratulating you on your new and wonderful position and on your promotion and outstanding success. Run this mental movie frequently, and the Cosmic Power within you will lead and guide you to a new position with wonderful opportunities for advancement."

He decided to do just what I recommended. He did not overcome his problems of resentment toward his boss and his relatives all at once; but whenever he thought of them or his boss, he silently blessed them by saying: *I release you; God be with you.* After a few weeks, he was completely free of his grudges, resentment, and hostility.

He now occupies an executive position in one of the leading organizations in Los Angeles, and he bubbles over with enthusiasm. He glows with the power, enthusiasm, and zeal of a transformed man. He has proved the age-old doctrine, "Change your thought and you change your future."

SUMMARY

Points to Recall

1. You mold, fashion, and create your own destiny. Your future is your present thought made manifest on the screen of space.
2. The first step in changing your mind, body, and environ-

ment is to believe that the Cosmic Healing Power which made your body can heal you and bring into your life all kinds of blessings. The Cosmic Power is all-wise.

3. You must assume responsibility for your future and cease blaming your in-laws, God, life, and the universe. You can instantly contact the Cosmic Power through the medium of your thought and make your future what you want it to be.

4. Take time out morning and evening to implant in your mind, by regular, systematic, and constructive thinking, the mental patterns of success, prosperity, abundance, right action, harmony, and full expression.

5. Your future is in your own hands for the simple reason that whatever you sow in your mind, you will reap on the screen of space. To think that your future is in the hands of God and not dependent on your own initiative is rank superstition. Your thought and your feeling create your destiny.

6. You can always plan for the future. If you are planning something in the future, you are planning it now—this instant—in your mind. Likewise, if you are fearing delays, obstacles, and impediments, you are thinking of these things now and creating the delays you fear. In the Cosmic Mind there is no time or space; all is the Eternal *now*. Plan something glorious and wonderful *now* in your mind.

7. Meet your challenges, problems, and difficulties boldly. Realize that the problem is there—but God is there, too; then every problem is Divinely outmatched. Stand up to the problem as a child of God and grapple with it courageously, and the Cosmic Power will instantly come to your aid. When you begin, God begins.

8. You predict your own future by the way in which you think, feel, imagine, act, and react.

9. You can follow all the rules, tenets, rituals, and ceremonies of a particular church and still be miserable, frustrated, and unhappy. You, and only you are responsible for the way you think and believe.
10. Change your thought-life and you change your destiny.

15

How to be serene in this changing world

Ralph Waldo Emerson said, "Nothing can bring you peace but the triumph of principles." Knowledge of the principles of your mind will add to your peace, poise, balance, and security.

For example, the engineer follows the principles of mathematics when constructing a bridge; he understands stresses, strains, and other complex scientific calculations based upon immutable laws which never fail. Likewise, the chemist and the physicist obey universal laws and principles underlying these forces.

You must discover the Cosmic laws and principles of your mind and live according to the principles of health, happiness, and serenity of mind. The law of your mind is referred to again and again in the greatest of all creative and dynamic books, the Bible, and that is the law of belief. . . . *As thou hast believed, so be it done unto thee* (Matthew 8:13).

To believe is to accept something as true, and whatever your conscious mind accepts as true—whether good or bad—your subconscious mind brings to pass. You *are* what you contemplate; you *are* that which you feel and believe yourself to be. Contemplate the following truth morning, noon, and night, and you will find peace in this changing world:

Finally, brethren, whatsoever things are true, whatsoever things are honest, whatsoever things are just, whatsoever things are pure, whatsoever things are lovely, whatsoever things are of good report; if there be any virtue, and if there be any praise, think on these things (Philippians 4:8).

How an Alcoholic Found Inner Peace and Freedom

Some months agc I interviewed an alcoholic whose wife had left him and whose two daughters no longer recognized him or spoke to him. He was in a state of deep depression and melancholia. I explained to him that his sincere desire to give up alcohol was the first step in healing, to which he readily agreed. The next step was to realize that there is a Cosmic Power within him which would take away all craving and would compel him to assert his freedom from the habit.

I suggested that he practice a simple mental technique several times daily by imagining that I was congratulating him on his freedom and believing in the picture of his mind. He continued to do this for about four weeks, three times daily for about five minutes each time.

Finally, the mental picture was objectified in his experience: a new habit-pattern had been established. He had gradually established the mental equivalent through thinking and feeling his freedom, until finally the mental picture was objectified. He said to me, "I have discovered the power of my own mind. I am free! I have a peace and serenity of mind which I never knew before, and I am grateful."

The Bible says, *Acquaint now thyself with him, and be at peace . . .* (Job 22:21). This man had acquainted himself with the power of his thought and imagination, to which the Cosmic Power within him responded. Peace comes to you as you discover the principles of your mind and apply them constructively in your life.

She Learned to Forget Bitterness and Found Serenity

Some months ago I visited a woman in a local hospital. She was suffering from acute high blood pressure and acute colitis. She began to recount all her former mistakes and errors, and she condemned herself and her stupidity. She said to me that she hated another girl who had undermined her in the office, and she was completely bogged down by destructive emotions which were snarled up in her subconscious mind.

I sat quietly with her for some time and then told her what William James, the father of American psychology, said. "The mark of genius is to know what to overlook." That is, you must turn away from the past—i.e., forgive and forget—if you want to be in perfect health and to have peace of mind.

I quoted a passage from the Bible and emphasized the fact that it was one of the greatest therapeutic agents available to bring about mental and physical health: *...Forgetting those things which are behind, and reaching forth unto those things which are before, I press toward the mark for the prize...* (Philippians 3:13-14). I explained to her, "The prize that you are seeking is peace of mind. If you have peace of mind, you also will have balance and equilibrium in your body. Peace means balance, equilibrium, equanimity, and serenity, which are due to a sense of oneness with God, the Infinite, the Whole of Life."

Jesus said to sick people: *...Thy faith hath made thee whole; go in peace* (Luke 8:48), because their minds were in turmoil and full of discord, confusion, inner warfare, and anger.

I explained to this woman that she could get back on the beam and find the peace that passeth understanding and that peace of mind is not escapism or a retreat from life. On the contrary, it is a constructive attitude of mind where you are vitally interested in the welfare of others and where you are a dynamic, creative person full of creativity, love, and goodwill to all.

I wrote out the following prayer for her to use frequently during the day:

I fully and freely forgive everyone who has ever hurt me. I release them, and it is done with forever. Whenever I think of any one of them, I bless that person. I forget the past and give my attention to a glorious future of perfect health, harmony, and peace. My mind is poised, serene, and calm. In this atmosphere of peace and goodwill which surrounds me, I feel a deep abiding strength and freedom from all fear. I now sense and feel the love and beauty of the Cosmic Healing Presence.

Day by day I become more aware of God's love; all that is false fades away. I now allow the Cosmic river of peace and healing power to flow through my whole body. I rest in the everlasting arms of peace. My peace is the deep unchanging Cosmic peace, the peace of God.

She practiced this prayer frequently during each day, and when I visited her again in about two weeks, she was very happy because she felt so much better; she was being released from the hospital the next day and had been pronounced healed. She said to me, "I know all my trouble was hate and resentment; I feel clean inside now. Truly, peace is health and happiness."

How an Executive Found True Serenity

Following a lecture in Honolulu a few years ago, an outstanding business executive asked me to have dinner with him that evening, as he had a very serious problem and thought I could help him. He was urbane, philosophical, and wealthy, and he seemed well-balanced and poised. But, in the intimacy of our conversation, he revealed that within he was a seething cauldron, full of hostility, resentment, and suppressed rage.

He asked me rather pathetically, "Can I have peace of mind? Serenity? And how can I get it? I want to sleep nights!"

Of course, I told him how to acquire a basic tranquility and an inner quietness. I continued, "Real power comes from inner quietness. The quiet and serene mind gets things done. God is peace and is at the very center of your being. The Bible says, *Peace I leave with you, my peace I give unto you: not as the world giveth, give I unto you. Let not your heart be troubled, neither let it be afraid.*[1] *Thou* (God, that is) *wilt keep him in perfect peace, whose mind is stayed on thee . . .*[2]

"In the midst of all your business difficulties and conflicts in daily life, keep your mind fixed on God with faith and confidence, and you will find His golden river of peace flowing through your mind and heart. Give the best that you have to your work; be creative and advance the good in life. Wish for all men what you wish for yourself. Contribute your talents and abilities to some good cause in your neighborhood. You are too absorbed with yourself.

"Commit to memory the passages I gave you, and every morning when you go to work close the door of your office and affirm these Cosmic truths. You will find their therapeutic, healing qualities flowing through your mind and through every atom of your being."

He said, "I feel better already! I now know what my problem is and what the answer is, and it is that I give peace to myself by staying my mind on God—and not on my problem!"

Last year when I visited Hawaii, I met him again, and he really is a new man. He has found the serenity of peace within himself. He has framed on the wall of his home this beautiful, soul-stirring, spiritual gem: *Thou wilt keep him in perfect peace, whose mind is stayed on thee . . .* (Isaiah 26:3).

How You Can Find Complete Serenity in Any Circumstance

Edwin Markham said, "At the heart of the cyclone tearing the sky, is a place of central calm." In the midst of a great hurricane or cyclone, the aviator knows that if he flies right into the center, there are stillness and calm. In the midst of you dwells the Cos-

[1] John 14:27.
[2] Isaiah 26:3.

mic Power which is all bliss, absolute peace, boundless love, absolute harmony, and all joy. Tune in mentally and emotionally with these qualities and attributes of Cosmic Wisdom and you will find yourself spiritually refreshed, replenished, and calm.

Your mind is a receptive medium for all the propaganda, opinions, and erroneous impressions coming from the world outside. Some of these impressions are good, but most are of a very negative nature. Except your mind is tuned in with the Cosmic Power and wisely sifts the chaff from the wheat, the negative and erroneous impressions take root and cause trouble, such as sickness, confusion, fear, and limitation of all kinds.

The world, or mass mind, believes in both good and evil powers, and in sickness, misery, misfortune, and catastrophes of all kinds. If you persevere in these worldly beliefs and neglect to tune in with the Cosmic Power, you shall have tribulations, trials, and difficulties.

The Bible says, . . . *Be of good cheer; I have overcome the world* (the mass mind) (John 16:33). Be of good cheer; let your knowledge of the Cosmic Power overcome all your problems. Begin now to fill your mind with concepts of harmony, peace, love, joy, and right action. Become aware of your inherent, God-given powers which enable you to give attention, devotion, and love to ideas which heal, bless, inspire, elevate, dignify, and fill your soul with joy.

You will always move in the direction of the idea which dominates your mind. Your awareness and confidence in the Cosmic Power and in Its Healing Presence which responds to you, fills your mind and heart and enables you to soar aloft over all obstacles to the haven of rest within yourself, where you may abide in the conviction that with the Cosmic Power of God all things are possible.

Maintaining this attitude of mind in the presence of difficulties will enable you to overcome the world (objective convictions and worldly fears and false beliefs). You shall be like the one spoken of by the Psalmist: *And he shall be like a tree planted by the rivers of water, that bringeth forth his fruit in his season; his leaf also shall not wither; and whatsoever he doeth shall prosper* (Psalm 1:3).

The Prayer That Gives You Serenity in This Changing World

Peace begins with me. The serenity and peace of Cosmic Power fills my mind; the spirit of goodwill goes forth from me to all mankind. I dwell in the secret place of the Most High. I claim sincerely and lovingly that all members of my family, all my associates, and all people everywhere are Divinely guided to their true expression in life where they are Divinely happy and prospered in all ways. The Cosmic river of peace moves through my mind and heart, and I radiate peace and goodwill to all people everywhere. I am always enveloped and surrounded by an infinite circle of right action and Divine love.

I affirm and boldly claim with confidence and faith that the wisdom of God in his Cosmic Wisdom anoints my intellect. I am inspired from On High. I see harmony where discord is; I see peace where pain is, love where hatred is, joy where sadness is, and life where so-called death is. All my loved ones and associates are included in my prayer, and the circle of God's love permeates their whole beings. If I have had trouble with another, I freely forgive him or her now. I completely let go of all bitterness and hostility. I see the image of God in others and wish for that person health, happiness, peace, and the blessings of the Eternal Cosmic Wisdom.

I give freely of my love, my wisdom, my understanding, and my wealth, distributing through Divine direction the riches of the Infinite to others. The peace of God which passeth all understanding fills my mind and heart, now and forevermore.

Change and decay, all around I see;
Oh, Thou who changeth not, abide with me.

SUMMARY

Highlights to Remember

1. Nothing can bring you peace but the triumph of Cosmic principles. Learn the Cosmic laws of your mind and experience peace, poise, balance, serenity and security.
2. You are what you contemplate. You are that which you believe yourself to be, for the law of mind is the law of belief.
3. You can imagine a friend or a loved one congratulating you on the answer to your prayer, and as you run your mental movie frequently and keep on believing in the picture of your mind, you will subjectify your mental image, and the answer comes forth.
4. You get acquainted with the Cosmic Power within you by using your thought and mental imagery, and the Power responds according to the nature of your request. You can contemplate freedom from alcoholism or from any other bad habit, and the Cosmic Power will free you from your craving and weakness.
5. *The mark of a genius is to know what to overlook.* You must forget the past errors, grievances and hurts and forgive yourself and others, if you want perfect health, happiness, and peace of mind.
6. Peace of mind is not escapism or retreating from life. On the contrary, it is a dynamic way of living, whereby you meet all obstacles head on and overcome them through

the Cosmic Power and wisdom of God. When you are creative, outgoing, doing what you love to do, contributing to the good and welfare of others, and full of love and goodwill to all, you have a dynamic sense of peace and inner quietude.

7. Real power comes from inner quietness. The quiet mind gets things done. God is absolute peace and dwells at the center of your being. Tune in on the God-Presence and let His river of peace flow through you, restoring your soul.
8. Contribute your talents and abilities to some good cause in your neighborhood and lift others up, and work for all that is constructive and which blesses mankind.
9. Stay your mind on the Cosmic Power which knows all and sees all, and which wipes away all tears and solves all problems. It is written: *Thou wilt keep him in perfect peace, whose mind is stayed on thee...* (Isaiah 26:3).
10. *At the heart of the cyclone tearing the sky is a place of central calm.* When troubled, perplexed, confused, or fearful, still the wheels of your mind and say over and over again, "Peace, be still," and you will find a great calm.
11. Remember, peace and serenity begins with you. You experience it by calling on Cosmic peace and love to fill your heart and mind, and you will experience the peace that passeth understanding.

16

How to overcome worry

Prolonged worry robs you of vitality, enthusiasm, and energy and leaves you a physical and mental wreck. Psychosomatic physicians point out that chronic worry is behind numerous diseases such as asthma, allergies, cardiac trouble, high blood pressure, and a host of other illnesses too numerous to mention.

The worried mind is confused, divided, and is thinking aimlessly about a lot of things which are not true. Worry can be solved, however, by the application of the law of mind.

Your problem is in your mind. You have a desire, the realization of which would solve your problem. But when you look at conditions and circumstances as they are, a negative thought comes to your mind and your desire is in conflict with your fear. Your worry is your mind's acceptance of the negative conditions.

Realize that your desire is the gift of God, telling you to rise higher in life and that there is no power to challenge God, the Living Spirit Almighty within you. Then affirm to yourself: *God (Cosmic Wisdom) gave me this desire, and the Almighty Power is now backing me up, revealing to me the perfect plan for its unfoldment, and I rest in that conviction.* When fear or worry thoughts come to your mind, remind yourself that God is bringing your desire, ideal, plan, or purpose to pass in Divine order, and continue in this attitude of mind until the day breaks and the shadows flee away.

How Business Worry Was Overcome

Recently I interviewed a businessman whose doctor told him that there was nothing wrong with him physically, but that he was suffering from "anxiety neurosis," which simply means chronic worry. This man told me, "Every time I pray or think about success, prosperity, and greater wealth, I start to worry about money, my business, and the future. It's wearing me down, and I am so tired." His vision of success and prosperity was thwarted by his chronic worry, and the fretting consumed his energy fruitlessly.

The way he overcame his anxiety neurosis was as follows: he began to have quiet sessions with himself three or four times a day when he declared solemnly:

> *But there is a spirit in man: And the inspiration of the Almighty giveth them understanding* (Job 32:8). *This Almighty Power is within me, enabling me to be, to do, and to have. This wisdom and power of the Almighty back me up and enable me to fulfill all my goals. I think about the wisdom and power of the Almighty regularly and systematically, and I no longer think about obstacles, delays, impediments, and failure. I know that thinking constantly along this line builds up my faith and confidence and increases my strength and poise, For God hath not given us the spirit of fear; but of power, and of love, and of a sound mind* (Timothy II 1:7).

In about a month's time this man arrived at that awareness of the strength, power, and intelligence which were Divinely implanted in him at his birth. He has conquered his worries by partaking of the spiritual medicine of Cosmic Wisdom.

How a Mother Dispelled Her Inner Fears

About a year ago, a distraught mother visited me, saying that she was terribly worried about her son in Vietnam. I gave her a specific prayer to use night and morning for herself and for her son. Subsequently, her son returned from Vietnam; he married and settled down, and she again came back to see me just as worried as before. I reminded her that her problem had been solved, but that she was still worried.

She was worried during the second visit that he may have married the wrong girl; however, she admitted that the girl was a wonderful wife. But she said, "All the time, I was so worried that their child might be born dead or crippled, but my daughter-in-law has given birth to a perfect child." The mother was then worried about a money shortage in her son's home!

This woman was not really worried about what she *thought* she was worried about. Her actual difficulty was that she had an inward sense of insecurity and was emotionally immature. While talking to her, I was able to show her that she was the creator of her own fears, and she thereupon replaced her inner sense of insecurity with a real feeling of security. I wrote out a special prayer for her to use.

> *He that dwelleth in the secret place of the most High shall abide under the shadow of the Almighty. I dwell in the secret place of the most High; this is my own mind. All the thoughts entertained by me conform to harmony, peace, and goodwill. My mind is the dwelling place of happiness, joy, and a deep sense of security. All the thoughts that enter my mind contribute to my joy, peace, and general welfare. I live, move, and have my being in the atmosphere of good fellowship, love, and unity.*
>
> *All the people that dwell in my mind are God's children. I am at peace in my mind with all the members of my family and with all mankind. The same good I wish for myself, I wish for my son and his family. I am living in the house of God now. I claim peace and happiness, for I know I dwell in the house of the Lord forever.*

She reiterated these truths frequently during the day, and these wonderful spiritual vibrations neutralized and obliterated the disease-soaked worry center in her subconscious mind. She discovered that there were spiritual reserves on which she could call to annihilate the negative thoughts. As she saturated her mind with these wonderful spiritual verities, she became possessed by a deep faith in all things good, and she is now living in the joyous expectancy of the best.

A Practical and Workable Approach to Overcoming Anxiety

A banker recently asked me to give him a simple, practical prayer to overcome his anxiety. I suggested that every morning before the day's work he be alone and identify himself mentally and emotionally with these truths:

> *I live, move, and have my being in God, and God lives, moves, and has His being in me. I am immersed in the Divine Cosmic Presence which sorrounds me, enfolds me, and enwraps me. My mind is God's mind and my spirit is the Spirit of God. This Infinite Being within me is the only Presence and the only Power; It cannot be defeated, thwarted, or frustrated in any way. It is All-Powerful and All-Wise, and It is present everywhere.*
>
> *As I unite mentally with this Infinite Power through my thoughts, I know I am greater than any problem. I grapple courageously with all difficulties and problems, knowing that they are Divinely outmatched, and whatever strength, power, and creative ideas I need will automatically be given to me by the Divine Cosmic Presence. I know the Infinite lies stretched in smiling repose within me where all is bliss, harmony, and peace. I am now in tune with the Infinite, and His wisdom, power, and intelligence become active and potent in*

my life. This is the law of my being, and I know God's peace fills my soul.

This banker understood what he was doing and why he was doing it. He said to me, "As I affirm these truths, I imagine they are falling down from my conscious to my subconscious mind and, by a process of spiritual osmosis, through every atom of my being." Every time I meet him in the bank, he tells me of some additional wonderful event in his life and those of his associates. He is calm, serene, poised, and balanced. He knows that the quiet mind gets things done.

How Fear of Auto Travel Was Conquered

A salesman had suffered a bad auto accident and was terribly worried every time he got into his car to drive through cities. I gave him a very simple technique, explaining to him that his mind could not contain two ideas simultaneously, i.e., he could not fear his journey and bless his trip at the same moment. Therefore, he had to supplant his worry with confidence and a sense of security. He began to bless his car as follows:

My car is God's car. It is God's idea and its creation came out of the One Cosmic Mind common to all individual men. This guides me and directs me in all my movements. Divine law and order govern me in my driving, and I go from town to town freely, joyously, and lovingly. I bless all other drivers on the road and wish for them all the blessings of life. I am an ambassador of God. I know that all the parts of my car are God's ideas and are functioning perfectly.

I am always poised, serene, and calm. I am always alert, alive, and quickened by the Holy Spirit. This love surrounds me and goes before me, making straight, joyous, and perfect my way. I am always surrounded by the sacred circle of love. It is wonderful!

During the past three years he has not had an accident, and he has received no citations or traffic violation tickets! He began to fill his mind with these truths and crowded out of his mind all worry and fear thoughts which had haunted him. He said, "I made it a habit to use that prayer all the time I was on the road. I committed the whole thing to memory and knew the higher vibration of my spiritual thoughts would wipe out the lower vibration." This salesman is no longer worried or fearful. He knows that prayer changes things.

Your Invisible Partner

I was in a drugstore in Detroit, and the pharmacist invited me to come back of the counter where he showed me a sign over the prescription department: *I will fear no evil: for thou art with me . . .* (Psalms 23:4). He added that his store had been robbed by gangsters three times and he had been held up twice with a gun pointed at his head. Following is the essence of his conversation:

"I think of that sentence of the Psalm, and it falls as a blessing on my mind. I have taken God as my partner, and I claim many times during the day: *God is my Higher Self, my Senior Partner. He guides me and watches over me. His power and wisdom are instantly available to me. I am not alone.* Now I feel secure because I know God's circle of love surrounds the store, myself, and all my customers."

This pharmacist met the problem of fear, anxiety, and worry, and he overcame it. During the past four years he has had no trouble and has prospered beyond his fondest dreams. He realized that his worry was irrational thought, and he became a straight-line thinker.

How a Schoolteacher Disposed of Worries

A schoolteacher told me that the way she overcomes all her worries is to take her worries apart; she holds them up to the light of reason, dissects them, and cuts them up into small pieces, and then she asks herself, "Are they real? Where do they come from? Do they have any power? Is there a principle behind them?"

With her cool, rational thought, she dismembers her worries and realizes they are shadows in her mind, fallacious and illusionary. Her final summation is, "Imagine an educated schoolteacher like myself worrying about shadows which have no reality!" She concludes by laughing her fears away.

Emotional Spasms Healed

I sent a friend of mine to see a heart specialist, as he was worried that he had a bad heart. The specialist took a cardiogram and told him that his heart was normal, and that his only problem was that he suffered from emotional spasms and seemed to be unreasonably obsessed with the idea that he had a bad heart.

The doctor told him that he should artfully and subtly instill in his mind the contents of the 27th Psalm until he let go of his false idea. In a few weeks' time he broke the spasmodic grip. He practiced the great law of substitution by repeating the good idea over and over again until the mind lays hold of the truth which sets you free and serene

How Worry of High Blood Pressure Was Handled

A short while ago a man with a seemingly well-adjusted and composed personality came to see me, very worried and anxious because his personal doctor had told him his blood pressure was over 200 and that he should take it easy and relax more. He said to me, "I can't take it easy! I have too much to do, and the pressure in my organization is terrific." He was really suffering from a long-mounting accumulation of petty frustrations and worries.

I expounded on the great truth that every fact of life is under the control and influence of the law of change. The old hymn says, "Change and decay all around I see; Oh, Thou Who changeth not, abide with me." God changes not and is the same yesterday, today, and forever, but all conditions, circumstances, and events are subject to alteration. Every created thing will someday pass away. The age-old maxim, "This too shall pass away," is always true.

I suggested that he begin to apply this truth to himself: that he could not be sick forever, that he was here to meet all problems and to overcome them, and that he was mentally and spiritually equipped to do so.

The first step was that he had to abstract his attention from his ailment and business difficulties and trust the Creative Cosmic Power within him, which made his body, to heal and to restore him. I gave him the following spiritual prescription to take in addition to the medicine which his doctor prescribed, accompanied with the suggestion that he was to assert absolutely and believe implicitly the following truths:

> *Periodically during the day I withdraw my attention from the vexations and strifes of the world, and I return to God and commune with Him. I know I am nourished spiritually and mentally, and God's peace floods my mind. God reveals to me the perfect solution and the perfect idea for every problem I meet. I reject the appearance of things, and I affirm the supremacy of the Infinite Power within me. I am absorbed and engrossed in the truth that God is guiding me and that Divine right action reigns supreme. The miraculous healing Cosmic Power is flowing through me and permeating every atom of my being. His river of peace flows through my mind and heart, and I am relaxed, poised, serene, and calm. I know the God-Presence which made me is now restoring me to wholeness and perfection, and I give thanks for the healing which is taking place right now.*

By affirming regularly many times a day in this manner, he succeeded in retrieving his senses from the annoyances and irritations of the day, and in a month's time a medical checkup revealed normal blood pressure. He discovered that his renewed mind restored his body to wholeness. When the strain and pressure of business tend to disturb him today, his motto is *None of these things moves me.*

As he exalts Cosmic Wisdom and God in the midst of him, his problems have become correspondingly smaller and smaller. Nobody rubs him the wrong way or excites him, and he feels adequate through the power of God to solve all problems and to meet all challenges. He has revalued himself in terms of his inner spiritual powers. *I will lift up mine eyes unto the hills, from whence cometh my help* (Psalms 121:1).

A Prayer to Banish Chronic Worry

Affirm daily with deep feeling:

> *I have a new, strong conviction of God's Cosmic Presence which holds me spellbound, entranced, and unmoved, and I feel serene, confident, and unafraid. I know there is nothing to fear—nothing to draw away from—for God is all there is and is everywhere present. In Him I live, move, and have my being; so I have no fear. God's envelope of love surrounds me and His golden river of peace flows through me; all is well.*
>
> *I am not afraid of people, conditions, events, or circumstances, for God is with me. Faith in God fills my soul, and I have no fear. I dwell in the Presence of God now and forevermore, and no fear can touch me. I am not afraid of the future, for God is with me. He is my dwelling place, and I am surrounded with the whole armor of God. God created me and He sustains me. God's wisdom leads and guides me, so I cannot err. I now celebrate the conviction of God's Presence because I know in my heart the great truth: "Closer is He than breathing, nearer than hands and feet."*

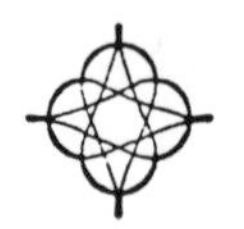

SUMMARY

Truths to Review Daily

1. Worry is having greater faith in your problem or negation than in God and His Cosmic Wisdom. Worry can be released by applying the Cosmic laws of mind.
2. Worry is caused by letting your desire or your ideal war with your fear thoughts. Realize that your desire is the Cosmic gift of God and that as you nourish it and sustain it with faith and confidence, you will bring it to pass.
3. Anxiety neurosis means chronic worry. Chronic worry comes from thinking confusedly about many things which are not true. Worry is a negative habit. Overcome your worry by realizing that the Power of the Almighty within you enables you to be, to do, and to have what you need. Dwell on this truth, and the more you do so, the more will your worries dwindle away.
4. You are not always worried over what you *think* you are worried about. Remember that chronic worry is basically due to a sense of insecurity, a feeling of alienation from God and his Cosmic goodness. Realize that you are the *creator* of your worries and that you can also supplant them with ideas of Cosmic Wisdom. This is the great law of substitution which frees you.
5. The Supreme Power within you cannot be defeated, vitiated, or thwarted in any way. It is all-powerful, and as you unite with this Presence and Cosmic Power through your thought, this Power will become active within you.

6. You can bless your car or any means of conveyance which you use for safe travel by realizing that Cosmic Wisdom goes before you, making straight, joyous, happy, and peaceful your way.
7. Realize that God is your partner and you will fear no evil. Know that His Overshadowing Presence and Cosmic Power always watch over you, and you will lead a charmed life.
8. Dissect your worries, hold them up to the light of Cosmic reason, and you will discover that they have no reality, and are but shadows in your mind, fallacious and illusory. Laugh your worries away and stop fighting shadows.
9. If obsessed with the idea of sickness or failure, realize you can release that mental grip by instilling artfully and persistently the great truths of the 27th Psalm, the greatest antidote to worry and fear in the world. By repetition and stick-to-it-iveness your mind will lay hold of the truth which sets you free.
10. A great truth is that everything in the world passes away. You can't have pain forever or be sick forever. Everything changes to its opposite "This too shall pass away"

17

How to link your thought to Cosmic Power

It is said that thought rules the world. Thought always finds its way into action. Ralph Waldo Emerson said, "Thought is the property of those only who can entertain it." He also said, "Man is what he thinks all day long."

Learn to respect and to have a wholesome regard for your thoughts. Your health, happiness, success in life, as well as your general peace of mind are largely determined by your awareness of the power of thought.

You have heard it said that thoughts are things, and that thoughts execute themselves. Your thought is a mental vibration and a definite power. Your action is but the outer and worldly manifestation and expression of your individual thought. If your thoughts are wise, your actions will be wise. When you conceive and ponder the thought, you are really releasing its latent power into action. William Shakespeare said, "Our thoughts are ours; their ends none of our own."

Think on whatsoever things are true, lovely, and elevating, and be assured of the authority of your thought. Believe in the Cosmic powers of your thought, and you will discover that thinking makes it so. Whatever you think and feel as true, that will you bring into your life. Your thought and feeling create your destiny.

Feeling—insofar as your thought is concerned—means *interest.* This is the meaning of the Biblical phrase, *As a man thinketh in his heart, so is he* (Proverbs 23:7). When you are vitally interested in your profession, work, or a special assignment, you will be successful because you have your heart in it. You are thinking in depth or feeling the reality of your thought, which is "thinking in the heart." Think with confidence in the power of your thought, and wonders will happen in your life.

Thinking in Cosmic Truth Brought Him Several Promotions

While I was talking to a detective last year, he told me that a case had baffled him for many months and that he had been thinking of a solution and guidance but that no answer came. I discovered that he believed it was a very difficult case, that the culprits had left the country, and that his search was thus more or less hopeless.

I told him the first thing he had to do was to change the belief or thought in his heart (his inner feeling nature); that actually he was double-minded; and that it was mandatory that he change his belief by realizing that Infinite Cosmic Intelligence in his subconscious is All-Wise and All-Knowing, and can reveal to him the whereabouts of the criminals who participated in the theft. He realized immediately that his prayer and thinking had been contradictory and that his head said one thing but that his heart said another; therefore, nothing happened. He used this formula:

> *I have a deep, abiding faith and conviction in the supremacy of the Infinite Intelligence within me. I am now turning over my request in this particular case to my Deeper Mind which is the Great Fabricator, and I know the answer will come automatically through my subconscious mind. I know that this belief in my heart*

is at this moment operating in Divine order, and I give thanks for the answer.

He turned over his request to his subconscious mind each night as if it were the first time he had done it, thereby continuously reinforcing his thought-patterns. At the end of one week, he had succeeded in penetrating the deeper layers of his subconscious mind, and he had reached a conviction.

On the morning of the eighth day, as he was shaving, the thought popped into his mind of a certain small town in Southern California. He went there with his partner and found the two men in a bar, arrested them, and recovered the stolen jewels. He had thereby proved without a doubt the wonders of thinking with Cosmic authority.

A Remarkable Case of a Man from India

Not long ago, while I was lecturing in Palm Springs, I met a man in the Spa Hotel who said, "Do you remember what you told me in India about sixteen years ago?"

I did not recall the man or our conversation; however, he reminded me of our conversation. He had wanted to come to America, but he was thinking of the difficulties of procuring a passport, the quota system, his financial destitution, and his lack of friends in the United States. I had told him that he was just thinking of obstacles, difficulties, and impediments, and reacting to environment and conditions. This was not true thinking; rather, it was foolish and absurd, because the mind magnifies what it looks upon.

He showed me the prayer that I had written for him over fifteen years ago:

Infinite Intelligence responds to my thought, for Its nature is responsiveness, and I know the way will open up for me to go to the United States in Divine Order. I think about this quietly and with interest until I get the response which satisfies me. As I think quietly, vividly, and lovingly on the answer to my prayer, I know I am activating the subjective wisdom in my subconscious mind; it then takes over and compels me to take all steps necessary for the realization of my desire.

He remarked that he had meditated along these lines morning and night for about a month, at the end of which time he had met an American industrialist for whom he had acted as a guide to many parts of India. The American, who was highly impressed with the man, arranged for his passage to New York and employed him as his private chauffeur at a wonderful salary.

This man knew what he was doing mentally, and his subconscious mind which received the impress of his thought had acted on the mind of the New York industrialist, causing him to fulfill the desire of this man in India.

You must remember that all men and women who seem to help you to attain your objectives are simply messengers employed by your Deeper Cosmic Mind to aid you in the unfoldment of the drama of your life.

How to Know if You Are Really Thinking with Cosmic Power

A man recently asked me: "How do I know when I am truly thinking?" That is a good question, and the answer I gave him is simply this: You are thinking when you are activating your mind from the standpoint of eternal verities and the Cosmic truths of God, which are the same yesterday, today, and forever. You are not thinking in the true sense of the term when you are reacting to the headlines in the newspapers, radio propaganda, or from the standpoint of tradition, dogma, creeds, or environmental conditions or circumstances.

This man who asked the question had been, as a matter of course, merely swallowing the thoughts of the columnists in the *Los Angeles Times,* and he was full of the ideas of local and state politicians. These were not his thoughts: rather, they were thoughts of others and were mostly of a negative nature.

He immediately saw the light and began to think for himself, taking as his spiritual yardstick:

> *Whatsoever things are true, whatsoever things are honest, whatsoever things are just, whatsoever things are pure, whatsoever things are lovely, whatsoever things are of good report; if there be any virtue, and if there be any praise, think on these things* (Philippians 4:8).

Whenever any thoughts or ideas were propounded to him, he would reason between the two opposing ideas or thoughts and come to a conclusion in his mind as to what was true from the standpoint of spiritual principles.

When the thought of failure came to him from time to time, he always affirmed boldly: *The Cosmic Infinite can't fail. I was born to succeed. Success is mine now, and the Infinite Intelligence in my subconscious mind responds and gives me the power, the strength, and the answer I need.*

He is now thinking from the standpoint of what is true of God, which is true Cosmic thinking. There is no longer any fear or worry in his thinking, and his mind is serene.

How a Mother Stopped Reacting to Negative Suggestions

A distraught mother recently came to see me, worrying that her son would get polio; also, the descriptions of cancer on television and the radio disturbed her, and she imagined that she had some of the symptoms about which she had learned. It seemed that every wind that blew affected her. She was not really thinking; she was reacting to the fears and negativity of the world.

I clarified in her mind the power of Cosmic thought, stating that every thought tends to manifest itself except it is neutralized by a counterthought of greater intensity and force, and that real spiritual thinking was completely free from fear and anxiety. She began to perceive that all her fears were but reactions to rumors, false beliefs, and an erroneous view of things. In other words, she was making externals the *cause*, instead of realizing that the Creative Cosmic Power was within her and would flow instantly in response to her directed thought.

She came to a simple forthright conclusion in her mind that hereafter she would separate the chaff from the wheat, the false from the true, fear from her ideal. She realized clearly that all her fears were due simply to a false viewpoint—i.e., giving power to externals and the false voices and propaganda of the world.

Recently this woman exclaimed, "It is wonderful to know that I am the only thinker in my universe, that *my* thought is creative, and that all the suggestions, propaganda, conditions, and circumstances of the world are not creative, but merely suggestive and all subject to change by *my* creative thought, and when I think God's thoughts, God's power is with my thoughts of good."

She is now a positive and constructive thinker because she thinks from the vantage point of Cosmic Truth.

A Healing Secured by Cosmic Thinking

I had a long talk with a woman who was quite bewildered and distraught by this passage in the Bible: *If any man come to me, and hate not his father, and mother, and wife, and children, and brethren, and sisters, yea, and his own life also, he cannot be my disciple* (Luke 14:26).

This woman had left her old traditional religion, which was that of her father and mother, and they were very critical of her; she deeply resented their gloomy pronouncement of eternal punishment for her.

The explanation was the cure for her mental disturbance. I pointed out to her that the word *hate* in the Bible means to reject, to repudiate, and to disabuse the mind of false concepts, and all that it meant was that she was to start thinking for herself from the standpoint of harmony, health, peace, wholeness, and all the blessings of life.

In other words, in spiritual Cosmic thinking there need be no antagonism or opposition to parents. They may have different religious beliefs and she was to permit them to believe what they wish and let them have their peculiarities, abnormalities, or eccentric beliefs, but she should love them in the sense that she wishes for them all the blessings of God, and she was to think kindly and lovingly of them at all times. She should, however, refuse and completely reject mentally the religious beliefs of her parents without any rancor or bitterness. I explained that Jesus was simply speaking in a parabolic way, exhorting men and women to follow wisdom and the eternal truths of life and to reject all erroneous beliefs and concepts about God, life, and the universe.

How She Gained Love and Goodwill

She gave up all her resentment and anger toward her parents and realized that as she worshipped, honored, and gave allegiance to the God-Presence within, she automatically had love and goodwill in her heart for her parents and for all men and women in the world. Love is the fulfilling of the Cosmic Law.

A Case of a Seemingly Virtuous Man Deprived of Good Fortune

A few months ago, I interviewed a man who stated, "I don't drink, smoke, gamble, or commit adultery; yet I am plagued by losses, misfortunes, and business failures, and I have been going to a psychiatrist for a mental disorder for over a year with no results. Why is God punishing me so?"

He was a member of a certain orthodox church; apparently he followed all the rules and tenets of his church meticulously from an external standpoint, and he prayed every night and morning and believed himself to be a good man, loyal and faithful to God. Now he was grumbling and growling against God and claiming that God was not playing fair with him!

I taught this man the simple workings of his own mind and

gave the reasons for all his troubles by elucidating on the fact that he could keep all the rules and regulations and practice all the formal rites of his religion and still bring all sorts of limitations, trials, tribulations, and sufferings on himself by wrong thinking and false beliefs.

His false concept that God (the Supreme Intelligence within him) was punishing him brought on all his difficulties by the very laws of his mind. . . . *As he thinketh in his heart, so is he* . . . (Proverbs 23:7). Actually he was tormenting and persecuting himself by his false beliefs. Moreover, he had a fear of failure which actually attracted failure in business and other ventures.

I also pointed out to him that it is not the body that commits adultery or remains virtuous. The body moves as it is moved by the mind. It acts as it is mentally acted upon. Your body cannot make you commit errors or steal, rob, or defraud. You mentally decree something and your body follows your mental orders. Your thought and feeling govern your life and are the causes of your experiences.

The interaction of your conscious and subconscious mind is the basis of all the events of your life. If you come to a false or an erroneous decision in your conscious mind, your subconscious mind accepts your decree—good, bad, or indifferent—and produces events and experiences according to your conscious mind belief.

This explanation appealed to him, and he began to use the following prayer regularly morning, noon, and prior to sleep at night:

> *I stay my mind upon God, and I am kept in perfect peace. In this peace of God I find order, harmony, and Divine love. No word of God shall be void of power. The words I speak are full of spirit and full of life. I acknowledge God in all my affairs and transactions. My mind is stayed on Him. I am blessed and prospered by the activity of the law of good operating in my life.*
>
> *All my statements of truth are charged with life, love, and meaning in Cosmic Wisdom. The words I speak*

make a deep impression on my Deeper Mind. I am glad that I know God dwells in my heart, for God is the very life of me. These words sink into my heart: "Behold, I dwell with thee, O man of God, and thou dwellest with me.

I know that when I look at my fellow man, I am seeing God in human form. I bless and send forth loving thoughts to all mankind. I speak the word now. My words are creative; the Deeper Mind within me responds. I now decree perfect health, harmony, and peace in my home, my heart, and in all my affairs.

I know and believe that God is guiding me now in all my ways and that Infinite Cosmic Spirit governs all my actions. I have decreed it; it shall come to pass. I have found the jewel of eternity in my own heart.

This man has received several promotions in his place of employment, a large industrial firm, and has found the peace that passeth understanding.

How to Advance by Beating the Law of Averages

A beautician asked me why she hadn't been getting ahead and making more money; she also complained that she felt inferior and unwanted, and she was really down on herself.

She was neither good nor bad, she earned a mediocre income, and was living in a backward area of the city. She was not thinking for herself, but rather was *producing and experiencing according to the law of averages.*

This law of averages is simply the mind of the masses which impinges on all of us and which automatically believes in sickness, accidents, disease, failure, and misfortunes of all kinds. The mass mind is mostly negative, but there is also much good in it, such as the constructive thinking of millions of spiritual minded people who pour into the mass mind thoughts of love, peace, joy, success, confidence, and faith in God and all things good; however, these people are outnumbered by the vast majority of people who dwell in negativity and ignorance. This is why no one should live solely according to the law of averages.

If you do not *think* for yourself and keep within Cosmic Consciousness you will automatically be a victim of the mass mind which impinges on the receptive medium of your mind; it will do all your thinking for you, resulting in negation and troubles of all kinds for you personally.

This young lady learned that what she was experiencing was the sum total of all she had ever thought plus the thoughts and beliefs of others, since she failed to think for herself. She began to activate her conscious mind spiritually, which soon became a law of action at her subconscious level. She began to discover a vast difference between spiritual (Cosmic) thinking and average (mass) thinking.

In average (mass) thinking, you are not in control of your thought life, and you are not giving correct commands to your subconscious. As she began her deliberate conscious mind selection of spiritual ideas, these automatically became impressed in her subconscious mind, and wonders began to happen in her life. She meditated on the following prayer several times a day:

> *I know and realize that God is a spirit moving within me. I know that God is a feeling or deep conviction of harmony, health, and peace within me; it is the movement of my own heart. The spirit or feeling of confidence and faith which now possesses me is the spirit of God and the action of God on the waters of my mind. This is God; it is the creative Power within me.*
>
> *I live, move, and have my being in the faith and confidence that goodness, truth, and beauty shall follow me all the days of my life. This faith in God and in all things good is omnipotent; it removes all barriers.*
>
> *I now close the door of the senses; I withdraw all attention from the world. I turn within to the One, the Beautiful, and the Good; here, I dwell with my Father beyond time and space; here, I live, move, and dwell in the shadow of the Almighty. I am free from all fear,*

from the verdict of the world, and the appearance of things. I now feel His Presence which is the feeling of the answered prayer, or the presence of my good.

I become that which I contemplate. I now feel that I am what I want to be; this feeling or awareness is the action of God in me; it is the Creative Power. I give thanks for the joy of the answered prayer, and I rest in the silence that "It is done."

This young lady is now the proud owner of her own beauty salon and is successfully moving onward.

SUMMARY

Highlights to Recall

1. Thought rules the world. Man is what he thinks all day long. Learn to respect and to have a wholesome regard for your thoughts and their power in your life.
2. Believe in the power of your thought, and you will discover that thinking makes it so.
3. Belief is one thing and thought is another. In order for spiritual thoughts to be 100 per cent effective, you must change your false beliefs about God (Cosmic Good) and His laws. If you *think* of a successful achievement, but *believe* you are going to fail, your belief of failure will be made manifest. Change your false belief and realize that you were born to succeed. Your thought of success will then become a reality. The Infinite within you cannot fail.

4. You are really thinking when there is no fear or worry in your thought. You are truly thinking when you realize there is an Infinite Cosmic Intelligence which responds to your thought. It is All-Powerful, All-Wise, and knows all. When your thought is clear on that, then the infinite powers of your subconscious mind will respond, giving you the answer. In spiritual thinking, you separate the chaff from the wheat, the fear from your desire, and you come to a definite conclusion that there is only one Power which responds to your thought.
5. Never give power to created things or to externals. In other words, never give power to any created thing. Give power only to the creative Power within you which responds to your thought. This banishes all fear.
6. You are not thinking when you are simply reacting to and agreeing with newspaper headlines, gossip, radio propaganda, and the opinions and false beliefs of others. You are thinking only when you think and interpret from the standpoint of God and His laws.
7. *You* are the only thinker in your universe, and you are responsible for the way in which you think. The suggestions and statements of others have no power, and you can reject everything unlike the truths of God. Reject everything that does not fill your soul with joy.
8. The word *hate* in the Bible means to disavow, to reject, and to repudiate all false concepts of God.
9. If you think that God is punishing you, you will simply inflict on yourself misery and suffering of all kinds because of your negative, destructive thinking.
10. The body does not commit adultery or steal or defraud. Your body cannot make you commit an error. You mentally think or decree a thing, and your body follows your mental orders.
11. The law of averages is the mass mind which is composed

of all manner of thoughts of the average mind, which believes mostly in sickness, misfortune, accidents, tragedies, and negation of all kinds. If you do not do your own thinking, this mass mind, or the law of averages, will do all your thinking for you and make a frightful mess of your life.

18

Let the Cosmic Power work wonders for you

The Cosmic law of your mind is an impersonal regulation and is no respecter of persons. When put into simple words, the law indicates that what you think, you create; what you feel, you attract; and what you imagine, you become.

The laws of mathematics, chemistry, physics, and electricity are equally impersonal and no respecter of persons. If you do not understand the laws of electricity regarding its conductivity, insulation, and the fact that it flows from a higher to a lowei potential, you could easily electrocute yourself. In other words, it is dangerous to meddle with forces you do not understand. The same rules would apply to the laws of chemistry; you should learn about atomic weights, the powers of valence, and the laws of attraction and repulsion, and as you make a deep study of organic and inorganic chemicals you would be able to blend marvelous compounds and bring forth new discoveries which would bless humanity in countless ways. If you make a mistake in addition, you will experience the results of that error.

Action and reaction are universal characteristics throughout all nature. Another way of stating this is to point out that any thought you feel as true is impressed on your subconscious mind (the law), and your subconscious expresses whatever is impressed upon it—good, bad, or indifferent.

If you think good, good follows. If you think evil, evil follows. For example, if you are opening a new business today, begin with faith and confidence that success is yours, because you were born to succeed. Claim that the wisdom, power, and intelligence of God are moving on your behalf, and the reaction will be that you will become a tremendous success and your business will go ahead by leaps and bounds.

The law is that the beginning and the end are the same. You began with the attitude of confidence, and your vision was on success, triumph, and accomplishment, and you experienced the automatic and mathematical response of the law, as the reaction is always the exact reproduction of your conscious thought and feeling.

How a Wife Reclaimed Her Husband's Love

A woman complained to me that after twenty years of a happy marriage her husband had gone "astray." During our conferences she mentioned that some months previously she had visited his office and had seen his new secretary, a beautiful blond, who is graceful, charming, and seductive; she admitted feeling pangs of jealousy and fear. I asked her if she had been fearing and picturing her husband as being interested in the girl, and she said, "Yes."

I elaborated on what she was doing and explained to her that her thoughts and the mental imagery of her husband straying had been conveyed subconsciously to her husband and they were impressed also on her own subconscious mind, as there is only one subjective mind, and what she feared most inevitably came upon her.

The intensity of her thoughts and the mental pictures were so strong that she actually accelerated her unfortunate marital condition. She had been using the law of her mind in a very negative way, and she had experienced the corresponding results.

Consequently, at my suggestion she talked over the whole situation with her husband and told him what she had been doing mentally. He in turn admitted his infidelity and decided to give up the other woman, and the bonds of love with God's blessings united them again. Her husband had known nothing about the laws of mind, but he is now studying *The Power of Your Subconscious Mind* [1] and applying it assiduously in his everyday life. Her success came from using the following scientifically worded prayer, night and morning:

> *I know that my husband is receptive to my constructive thought and imagery. I claim, feel, and know that at the center of his being is peace. My husband is Divinely guided in all ways. He is a channel for the Divine. God's Cosmic Love fills his mind and heart. There are harmony, peace, love, and understanding between us. I picture him as happy, healthy, joyous, loving, and prosperous. I surround and enfold him with the Cosmic circle of God's love which is impregnable, impervious, and invulnerable to all negation.*

Following the constructive use of the Cosmic law, her mind ceased its turmoil and resumed its natural state of peace, and the sacred bonds of marriage now reign supreme between husband and wife.

He Said, "This Is My Fifth Divorce! What's Wrong?"

A man, apparently in a state of desperation, asked me to throw light upon his marital problems because, as he said, "I have married five times successively hoping to get a good wife—and now I want a divorce from my present wife."

I thereupon emphasized the Cosmic law of mind and pointed out that the law of his mind is absolutely just and eminently fair

[1] *The Power of Your Subconscious Mind,* Joseph Murphy, Prentice-Hall, Inc., Englewood Cliffs, New Jersey, 1963.

in its manifestations, e.g., it is the nature of an apple seed to bring forth an apple tree; likewise, it is the law of life that man invariably and inevitably reproduces in all phases of his life the exact duplicate of his inner nature. "As within, so without. As in heaven (mind), so on earth (body, circumstances, conditions, experiences, and events)."

I believe I cleared up the obscurities in his mind through clarification of the way his thoughts and feelings operate in his life. He came to perceive clearly that it is impossible for a man to think, feel, and believe one thing and then to experience something other than what he thinks and feels to be true.

I told him that if you picture failure and feel yourself defeated, you can't possibly express success and triumph over obstacles. The law is considered not only good but very good because your experiences dovetail and correlate with your inner attitudes and beliefs.

He said to me, "I see now that divorce is not the answer as it would be no solution for me as long as I continue the patterns of jealousy, fear, anger, and possessiveness. I must change myself. I have been accusing my wife falsely and now I know that my imputation of wrong doing and recriminations were projections of my own guilt, fear, and insecurity."

He had come to a wise deduction, because he could have divorced his present wife and married another, but he would have repeated the same dreary pattern of jealousy, accusation, self-pity, depression, and suppressed rage. He began to affirm frequently during the day:

> *I know I cannot think of two things at the same time. I know that two objects cannot occupy the same space at the same time. I can't have thoughts of love and resentment at the same time. Whenever I think of my wife I state positively, "God's love fills her soul. I radiate peace, harmony, and goodwill to her. Our marriage is a spiritual union. I am one with God and with all people." I know I can make my wife's life full,*

complete, and wonderful. Only that which belongs to love, truth, and wholeness can enter our experience.

His wife didn't really want a divorce, but she was tired of his false allegations and invectives. She was delighted to hear that he had become interested in the science of his mind and was enraptured over his spiritual transformation. They are both using the law now in the right way.

Why a Manuscript Was Not Published

I interviewed a man in Hawaii last year who showed me a manuscript which I read and was very much impressed with. However, he had received twelve rejection notices from various publishing houses. Now, this was a perfect working of the law of his mind, as our successes or failures, our soundness of body or our afflictions, are the exact manifestation of our states of consciousness, which is the way we think, feel, believe, and whatever we give mental consent to.

This man had a picture of rejection in his mind. His thought was action, and his experience was the reaction. The action and the reaction corresponded. He reversed his mental attitude and affirmed boldly:

Infinite Cosmic Intelligence reveals to me the ideal publisher who will accept my manuscript, publish it, and promote it in Divine order. I accept this idea completely in my mind. I know that my mind is like my motion picture projector: if I don't like the picture reflected on the screen, I can change the film and project a new scene. Likewise, I know that the contents of my mentality are always dramatized and portrayed in my objective world. I know that my thought and feeling predict my future. I now enthrone in my mind constructive thoughts of harmony, Divine Guidance, and right action. I am conscious of my true worth, and I have a deep, abiding respect for the Divinity within which shapes my ends. I know that the infinite intelligence of my subconscious mind responds to my habitual thinking, and whenever fear thoughts come into my mind, I claim silently, "It is God in action in all phases of my life, bringing all my desires to pass at the right time and in the right way."

Following this Cosmic reordering of his mind, he was guided to find a publishing house which accepted and published his manuscript, and it is a big success. He is now writing his second book.

She Believed the "Cards Were Stacked" Against Her Happiness

I had an interesting interview with a woman in Los Angeles. She was present at a lecture I gave on *The Amazing Laws of Cosmic Mind Power*[2] at the Wilshire Ebell Theatre. She said, "My life seems to be predetermined by fate. I seem to be under a fixed sentence by which the order of things is prescribed like the stacking of a deck of cards. In the past four years I have been engaged to four men, and each year one of them has met an untimely death just prior to the marriage ceremony."

I asked her if she had ever read or heard of Ralph Waldo Emerson's definition of *fate,* and her answer was "no." *He* [man] *thinks his fate alien because the copula (link or connection) is hidden. But the soul [subconscious mind] contains the event that shall befall it, for the event is only the actualization of its thoughts, and what we pray to ourselves for is always granted. The event is the print of your form. It fits you like your skin!*

I explained to her that what she is within herself in terms of training, conditioning, theological concepts, emotional acceptance, thoughts, feelings, and beliefs, determines her conditions, experiences, and events; that her subconscious mind is always

[2] *The Amazing Laws of Cosmic Mind Power,* Joseph Murphy, Parker Publishing Company, Inc., West Nyack, N. Y., 1965.

reproducing her habitual thinking and beliefs; and that in order to change her life, she would have to change her thoughts and keep them changed. I elaborated on the fact that, after all, she did not control the lives of other men and women, and that if her fiancé, acquaintances, mother and father, or friends die or suddenly pass on to the next dimension of life by an accident or by another tragedy, she was not to blame, as each person demonstrates and manifests his own state of consciousness. *Actually, she had been constantly fearing consciously and subconsciously that whatever man she would attract would meet the same fate as her first fiancé.* She said to me, "I *knew* it would happen each time. It's Kismet."

She began to comprehend the law of her mind and what she had been doing to herself. Job truly said. . . . *the thing which I greatly feared is come upon me* . . . (Job 3:25). This motion or action of her mind had taken away from her the sense of happiness, well being, and fulfillment of her dreams, and she had attracted men whose consciousness had reached the point of transition to the next dimension of life. She reversed her mental attitude and affirmed boldly:

> *I know that two unlike things repel each other. I walk and talk with God, and I believe that God is guiding me and that the law of harmony is always governing me. I know that discord and harmony do not dwell together; I know that I can't cry and laugh at the same time. I know that when I claim, feel, and believe that God loves me and cares for me and that He guides and directs me, I can't be on a train that is wrecked because Divine law and order govern my life. Likewise, as I walk in the light of God's love, knowing that I am attracting a wonderful man who harmonizes with me perfectly spiritually, mentally, and physically, I know the law responds accordingly. Having met him* within, *I must meet him in the* without, *for this is the law of my mind. I know it is accomplished in Divine Mind now.*

At the end of a few weeks of meditating with Cosmic Power along the above lines, she met a dentist who subsequently proposed to her, and I had the joy and pleasure of performing the marriage ceremony. She had replaced her fear with faith and confidence, and she has discovered that wonders happen when you use the Cosmic law righteously.

He Prayed and Worked Hard, but Didn't Prosper

A young executive complained that he regularly prayed, using a formula in *Your Infinite Power to Be Rich,*[3] and that he believed what it said, but that he got no results. *This young man's trouble was that he was stealing from himself.* He admitted to me that he was envious of one of his associate's promotion and increased salary, and he also coveted his boss's immense wealth, mental acumen, and sagacity. He listened avidly as I informed him about the Cosmic Law that no one can break.

Cosmic Law is an infallible, immutable, changeless, and eternal law within man's heart (subconscious mind). Every thought is the beginning of action and tends to manifest itself, evoking a corresponding reaction from the subconscious mind. The Bible says in the 20th chapter of Exodus, *Thou shalt not covet . . .* which means that in his case he was envious inordinately and wrongfully of that which belonged to others. This was a mood of loss, lack, and limitation within himself and according to the law of mind it attracted more loss, such as loss of prestige, vitality, and promotion, and in general he was impoverishing himself.

I pointed out to him that he could attain nothing in life without establishing the mental equivalent in his mentality and that there is no such thing as a "free lunch" or something for nothing—but that he must claim, feel, and mentally accept his good to the point of his own conviction, and only then will his good automatically follow.

He realized that he had been stealing from himself wealth health, peace, and prosperity by his own thoughts of envy,

[3] *Your Infinite Power to Be Rich,* Joseph Murphy, Parker Publishing Company, Inc., West Nyack, N. Y., 1966.

jealousy, and covetousness. His thoughts were correspondingly made manifest in his job, his pocketbook, and in all phases of his life.

Joyful satisfaction came to him as he prayed regularly, systematically, and judiciously in the following manner:

> *I sincerely wish success, happiness, peace, wealth, and prosperity for all my associates and for all men everywhere. I rejoice in their promotion, success, and advancement. I now dwell on the Omnipresence and the Omniaction of God. I know that this Infinite Cosmic Wisdom guides the planets on their courses. I know this same Divine Intelligence governs and directs all my affairs. I claim and believe that Divine understanding is mine at all times. I know that all my activities are controlled by this indwelling Cosmic Presence. God's wisdom, truth, and beauty are being expressed by me at all times. The All-Knowing One within me knows what to do and how to do it with Cosmic Power. My business or profession is completely controlled, governed, and directed by It. Divine guidance is mine.*

Whenever the thought of envy, jealousy, or covetousness came to his mind, he affirmed: *I wish for him (or her) all the blessings of life.* After awhile the negative thoughts lost their momentum, and he reconditioned his mind to let Cosmic Power work for him. He has since won a marvelous promotion.

He Wanted to Outlaw War by Passing a Law

While I was giving a series of lectures recently in Palm Springs, a man in the hotel where I was staying asked me to sign a paper outlawing war. He said that he had 20,000 signatures so far and that he expected to obtain millions of signatures; then he was going to submit them to Congress and insist that they pass a law outlawing war and influence other nations to do the same. All this is so much balderdash and folderal.

In our conversation, I explained to him that men could sign all the documents for peace in every parliament in the world, and it would be of no avail. History shows and proves that many governmental decrees and formal agreements and pledges for peace have been signed by numerous nations but, often before the ink wherewith they were written could dry, they were broken. Parliaments and legislatures of the world cannot legislate peace, harmony, security, abundance, or love of neighbor—all these are ordained and legislated in the minds and hearts of man. Peace begins with the individual, and if a man has peace within himself he will be at peace with his wife, his friends and associates, as well as with all the world.

If a man is full of anger, resentment, hostility, and suppressed rage, he is at war with himself and his world. A nation is an aggregation of individuals; therefore, the only place to write legislation for peace is by the individual tuning in on the God of peace within himself and feeling that river of peace, love, harmony, and joy flowing through him. Furthermore, when man realizes that he can go to the Infinite Mind within himself and there claim and feel what he wants to be, to do, and to have, the Infinite Intelligence will respond accordingly, and he will discover that he can have what he wants without hurting the hair of a living being.

Furthermore, I added that war is caused by fear, hate, greed, revenge, anger, and the lust of men, and that this is why man's inhumanity to man makes countless thousands mourn.

Why Doesn't God Stop Wars if God Is Love?

This was another question propounded at one of my lectures. One woman said, "If God is love and if He is All-Good and All-Wise, why doesn't He stop war, crime, murder, and rapine? Why does He permit thousands and millions of children to die of hunger and countless other thousands to become crippled and maimed by the ravages of war?" She seemed to be angry at God.

The answer is very simple. God is the Universal Being, the Cosmic Power. the Supreme and Infinite Intelligence working on a cosmic scale or from the standpoint of the Universal. The law, simply stated, is that the Universal can only act on the plane of the particular or individual by becoming the individual. In other words, God rules the world and acts solely on the Cosmic plane, moving as unity, harmony, rhythm, order, beauty, and proportion. The only way God can work through you is through your thought, feeling and mental imagery.

You have volition, choice, and initiative. You have freedom to become a murderer or a holy man; otherwise, you would not be an individual. You are not compelled to be good or holy; you have freedom to choose harmony, peace, joy, love, abundance, and all the blessings of life.

You were not compelled to love your husband or wife. You said, "I choose him (or her) from all the others in the world." *... Choose you this day whom ye will serve ...* (Joshua 24:15).

As long as man remains emotionally immature, and while he harbors resentment, ill will, jealousy, hate, and anger, he is at war with himself and with others; multiply the one man and you have a nation.

The law of God governs the individual, the nation, and the world. God can't stop wars, crime, disease, discord, and accidents. All judgment is given to the son, which means your mind. You judge yourself by the way you think.

You are the one to *initiate* peace in your own mind and heart, and then *your* world will be at peace. There is no one to change but yourself. Start now:

> *... Whatsoever things are true, whatsoever things are honest, whatsoever things are just, whatsoever things are pure, whatsoever things are lovely, whatsoever things are of good report; if there be any virtue, and if there be any praise, think on these things.* (Philippians 4:8).

As you do this, your whole world will magically melt into the image and likeness of your concept of yourself, and your desert will rejoice and blossom as the rose. This indeed is letting Cosmic Power work wonders for you.

SUMMARY

Points to Recall

1. The Cosmic law of your mind is impersonal and is no respecter of persons. You can't break the law, for it is written in your heart and inscribed in the inward parts of your being. The law is: that which you think, you create; what you feel, you attract; and what you imagine, you become.
2. The laws of mathematics, physics, chemistry, and electricity are not different from the laws operating in your own mind. If you misuse or misdirect the principles of electricity, you will get into trouble. If you misuse or misdirect your mind by negative thinking, you will also get into trouble.
3. Action and reaction are universal throughout all nature. Your thought is action, and the reaction is the response of your subconscious to your thought life. Thought and its manifestation are one. You can't think one thing and produce another. Your thought and your experience are one.
4. A wife who pictures her husband as being unfaithful and who constantly thinks along these lines may literally experience what she fears, thus fulfilling the law of her mind and what she fears most comes upon her.
5. You can't express success, achievement, and victory by having a mental picture of failure. Your image of failure will be executed on the screen of space.

6. Two objects can't occupy the same space at the same time. Likewise, you can't think of success and failure at the same time. Think success and focus the lens of your mind on success and all the reasons why you can succeed, and you will succeed.
7. If you have a mental image of rejection in your mind, you will experience the same. As within, so without.
8. The cards are not "stacked" against you. Your thought and feeling create your destiny. Your future is your present thinking made manifest with Cosmic Power.
9. Two unlike things repel each other. If you believe that God is guiding you and that the law of harmony is always governing you, you can't be hurt on a conveyance that gets wrecked.
10. To be envious and covetous of others is to impoverish yourself and attract to you lack and limitation of all kinds.
11. No government or parliament can legislate peace, harmony, love, prosperity, or security. These are ordained in the minds and hearts of all individuals as they tune in on the sources of Cosmic Power.
12. God can't outlaw war, crime, or sickness because God is Universal Mind, the Cosmic Being operating on the plane of the Universal only, and God works and functions through your thoughts, feelings, and imagery.
13. God is not a man. God is Cosmic and Universal Spirit. You are an individualized expression of God. You have volition, choice, and initiative and also the inner knowledge of using the Cosmic Power, ever available for your supreme good in all things.